CW0068434B

NARROW BOAT ENGINE
MAINTENANCE AND REPAIR

NARROW BOAT ENGINE
MAINTENANCE AND REPAIR

Stephanie L. Horton BEng, CEng, MIET

THE CROWOOD PRESS

First published in 2017 by
The Crowood Press Ltd
Ramsbury, Marlborough
Wiltshire SN8 2HR

www.crowood.com

British Library Cataloguing-in-Publication Data
A catalogue record for this book is available from the British Library.

ISBN 978 1 78500 349 3

Typeset by Jean Cussons Typesetting, Diss, Norfolk

Printed and bound in India by Parksons Graphics

CONTENTS

PREFACE AND ACKNOWLEDGEMENTS

This book is based on the engine maintenance courses that River Canal Rescue has been running for many years, which were designed to teach the background theory and give the opportunity to gain practical experience. These were initially developed and presented by Tony Brooks, whose course manual has been the foundation of this book.

Over the years the course has developed further and more practical elements have been included to respond to boaters' suggestions and needs. I have tried to reflect these elements within the book, keeping to practical and useful knowledge, rather than going in to technical detail or explaining how to undertake work that only one person in a hundred may find useful, which in my opinion is best left to the companies and individuals that offer these services.

The book would not be as comprehensive as it is, or contain so many tips, without the input of the many engineers who operate daily on the inland waterways. Their guidance and help have proved invaluable when trying to work out the simplest way to explain some of the more difficult jobs.

Huge thanks to Tony Brooks for letting me use his original works and diagrams, and to Steve Beck and Jay Forman for proof-reading the book and not being afraid of telling me when I had got something wrong. A special mention to Keith Meadowcroft for his contribution to the electrical section, and to Jay Forman for coming up with practical ways of fault-finding on electrical systems.

To all of the engineers whose 'hands' appear in the photos, thank you for being patient, especially when we had to redo some of the servicing photos – I know that it's not your favourite job, but at least it was not out in the rain! Thank you also to everyone who had input or was used as a bouncing board when I needed clarification, your help and guidance is appreciated.

Finally a special mention to my husband Trevor Forman, whose engineering knowledge and skills were key to setting up River Canal Rescue: as he always reminds me, none of this would have been possible without him.

I hope that you find the book useful and practical and get some real benefit from the purchase.

INTRODUCTION

This book has been written to provide a practical guide to maintaining your diesel engine and its associated systems, along with the background theory to develop knowledge and understanding. There are tips and easy to follow guides to take you through each maintenance section. This book should enable a complete novice to gain understanding and build confidence in diesel engine and boat maintenance.

The book concentrates on the maintenance of the diesel engine, but also covers gearbox and drive systems, electrical systems and cooling systems, as it is important for every boat owner to develop a rounded knowledge of all systems and to feel confident in undertaking routine maintenance.

However, before we start it is important that you take the time to identify the type of engine, drive components and electrical system that are installed in your vessel. There is an Engine Details sheet included in the Appendix for you to note down the engine and component specifics for easy reference. Once you have these details, the information in this book will be easier to adopt and apply to your specific application.

This introductory chapter provides an overview of the systems we will be covering in the book and how they work as a whole. It will provide some of the background information that will assist in understanding the type of engine you have and

make you aware of the key benefits and issues that each may have.

There are many different types of engines installed in boats and, depending on the application, age of the vessel and usage, these engines can be specifically for marine applications, or vehicle and industrial engines that have been marinized for use in boats. However, all marine diesel engine use a number of similar components and therefore this book will provide you with a broad knowledge of these components and how they are integrated, ensuring that no matter what engine, gearbox or age of vessel, you should be able to apply the knowledge gained through this book to maintain the systems. The following section will assist you in identifying and classifying the type of engine and specific components installed in your vessel.

NOTE: The sections detailing the theory behind 'how it works' are provided for those wishing to explore the subject in a little more depth. Just knowing that an engine is a device for turning fuel into rotary motion is sufficient for those wishing to skip forward to the maintenance sections.

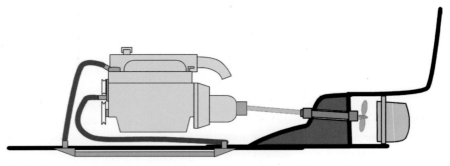

Typical arrangement of the engine, gearbox and propulsion systems.

THE DIESEL ENGINE OPERATION

The diesel engine is similar to the petrol engine, in that it operates on a four-stroke cycle. However, it is also very different, in that it has no ignition system, so there is no need for a spark plug to ignite the fuel. Instead, the air is compressed in the cylinder to increase the pressure rapidly; this in turn creates the high temperatures required so that as fuel is added, it causes the fuel to ignite. Other than a starter motor, the diesel engine does not require any further electrical interaction to operate. It is worth noting that diesel ignites at much higher temperatures than petrol, well over 300°C, which makes it much safer to use and store than petrol and is one of the main contributing factors for the move towards diesel engines on boats.

The four-stroke cycle describes the internal sequence of events that the engine goes through in order to produce power. It is commonly referred to as the 'suck, squeeze, bang and blow' cycle.

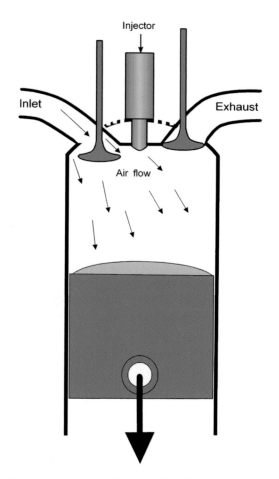

Piston cycle – induction ('suck'). The air is drawn in to the cylinder as the piston moves down; the inlet valve is open to allow air to enter. This is called the 'induction stroke'.

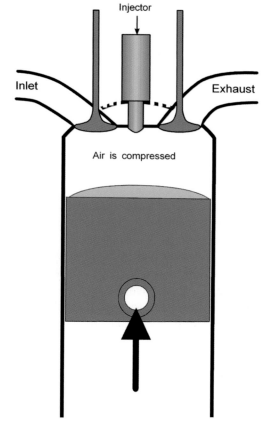

Piston cycle – compression ('squeeze'). The inlet valve closes and as the piston moves up the air is compressed (the air temperature increases as the air is compressed). This is called the 'compression stroke'.

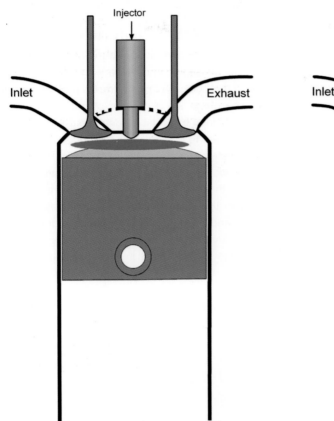

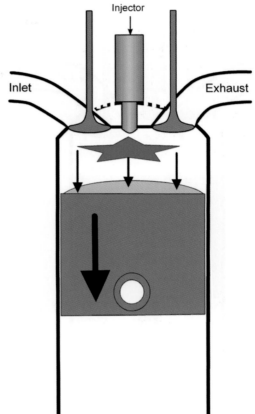

Piston cycle. The diesel is injected into the cylinder.

Piston cycle – ignition ('bang'). The diesel is injected into the cylinder as the piston reaches the top of its stroke. The compressed air is at its hottest point, commonly known as TDC (Top Dead Centre) and ignition occurs. As the diesel ignites and expands, it forces the piston back down. This is the 'power stroke'.

In order for a diesel engine to start, a number of conditions must be met. The engine must be turning over quickly enough to build the pressure to create the temperatures required for ignition. Flat batteries or poor connections will result in non-starting as the engine cannot be turned over quickly enough to create the ideal conditions for combustion.

Fuel is a vital component and must be added to the engine at exactly the right time and in the right amounts; too much or too little will result in irregular combustion and erratic running. This can be due to many causes, such as fuel contamination, injection pump failure, injectors and fuel pump issues.

Finally, the engine compression plays a big part in the success of the cycle. In order to produce the pressure in the cylinders, initially air must be drawn in at the right time. When the valve closes and seals, piston rings create a sealed surface in the cylinders so that the ideal conditions are met for compression to be achieved. When compression starts to fail, the piston rings or valve seals are wearing and therefore gases escape during combustion, causing poor or erratic running. This usually requires an engine strip and new parts to be fitted.

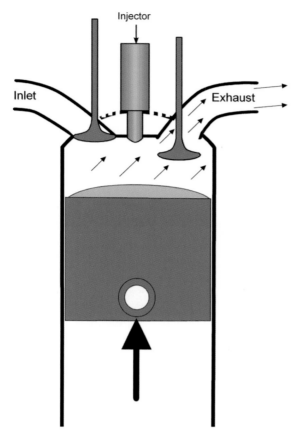

Piston cycle – exhaust ('blow'). The piston rises to push out the exhaust gases, and with the exhaust value open the air is expelled through the exhaust. This is called the 'exhaust stroke'.

TIP: TROUBLE STARTING

Any diesel engine will always start as long as it is spun over rapidly enough and the correct amount of fuel is added at the right time, even where compression is an issue, although you may have to use an additive like Easy Start to give an extra boost. (This is not recommended other than in an emergency, as an engine can become 'addicted' to this type of substance and any damage caused will be irreversible.)

MARINE ENGINE TYPES, SYSTEMS AND COMPONENTS

The photographs here show some of the main components on the engine and these will be discussed in detail in the book. The photographs provide a quick reference (there are a number of different engines detailed below) to assist in identifying what to look for and where to look on your engine.

The system in your vessel will consist of several different components that make it operational and each of these is covered in detail in this book. However, the following sections will help you to identify and recognize the important components on your engine and these should be recorded on the Engine Details sheet in the Appendix for easy reference.

Marine Engines

Bukh, Volvo Penta and MerCruiser, to name a few.

Volvo Penta engine, showing the left side and all the main components fitted to the engine.

These are engines specifically designed for marine applications. They are usually direct injected (*see* section below for explanation), heavy and comparatively slow revving. They are designed to operate at a coolant temperature of 70°C or less.

Automotive Engines

Typified by BMC 1.5 and 1.8, Perkins, Peugeot, Ford.

Typically four or more cylinders, these are usually high revving (for their age and design) and comparatively highly stressed. In normal marine uses they are usually indirect injected (*see* section below for explanation).

The cooling systems will normally operate at about 85°C and will have a pressurized system, unless derated by the supplier for direct cooling (*see* later section). Parts are relatively cheap and, with the right choice of engine, readily available.

Volvo Penta engine, showing the right side and all the main components fitted to the engine.

BMC engine, showing the left side and all the main components fitted to the engine.

BMC engine, showing the front and all the main components fitted to the engine.

Industrial Engines

Kubota (Beta and Nanni), Mitsubishi (Vetus and Thorneycroft), Isuzu (HMI), Yanmar (Barrus Shire), Lister.

The industrial engine is one of the more popular engines and the names above are the companies that have marinized the engine for marine applications. A reliable base engine, such as Mitsubishi or Kubota, is the most common basis and the indirect-injection system is usually employed. However, depending upon the age of the engine, its characteristics will vary in terms of design, stress and so on. Modern Japanese-based units have characteristics similar to automotive engines, whilst older Listers have more in common with marine engines.

BMC engine, showing the right side and all the main components fitted to the engine.

Vetus engine, showing the left side and all the main components fitted to the engine.

Vetus engine, showing the front and all the main components fitted to the engine.

Vetus engine, showing the right side and all the main components fitted to the engine.

Lister engine, showing the left side and all the main components fitted to the engine.

Lister engine, showing the right side and all the main components fitted to the engine.

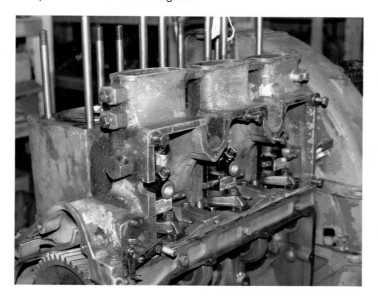

Classic/Vintage Engines
Lister, Gardner, Ruston Hornsby, Kelvin, Petter, Stuart Turner.

These tend to be engines over fifty years old, or old designs of engines from the late industrial era. They will normally be direct injected and they tend to be big, heavy and slow revving, with a very large torque. This slow revving, high-torque characteristic makes these engines ideal for marine applications and results in a long life.

DIRECT AND INDIRECT DIESEL INJECTION ENGINES

Whether an engine is designed specifically for marine applications or has been marinized to suit, all diesel engines fall into just two common designs that describe the method of diesel injection into the combustion chambers of the engine.

In direct-injection engines the fuel is sprayed directly into the combustion chamber. The top of the piston is shaped to direct and control the air movement, resulting in the fuel mixing with the heated air and burning more evenly. The engine benefits from good fuel economy and easy starting.

In an indirect-injection engine, the design incorporates a 'pre-combustion chamber', where the fuel is injected to mix with the heated air that is forced into it by the piston on the compression stroke from the previous cycle.

Lister engine, showing the internal injection pumps; there is one per cylinder. To access them, the side plate has to be removed.

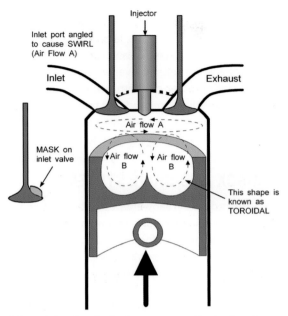

Direct injection (cylinder cross section), showing how fuel is injected and ignited.

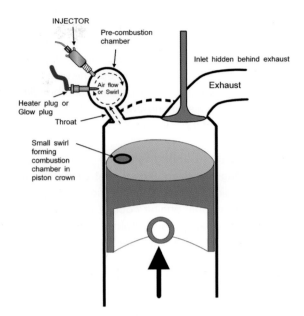

Indirect injection (cylinder cross section), showing how fuel is injected and ignited.

However, this design does require heater plugs (glow plugs) to assist with cold starting. The piston has a small 'spherical chamber', which causes the air and fuel mixture to swirl in the combustion chamber, and these two elements combined speed up the combustion process. Indirect injection benefits include longer engine life and smoother running, enabling higher revs per minute.

Most current engines, other than vintage engines, use the indirect-injection method; however, there is a general move towards direct injection with newer industrial engines. It is important to understand the differences and to identify which type you have. The easiest way to is to check if your engine uses heater plugs (glow plugs).

COOLING SYSTEMS

The cooling system on boats will vary depending upon the type of vessel and the engine; if you have an air-cooled engine (old Listers or non-marinized engines), then you will not need to spend time on this chapter. However, all other systems will fall into three categories: keel cooled; heat-exchanger cooled; or raw-water cooled. In some cases, a mixture of these cooling systems is in operation, if the vessel has been converted from one type to another. The following will assist in identifying which system you have.

Keel-Cooled Systems

Keel-cooled systems have tanks welded into the hull, or have pipes running on the outside of the hull to create a sealed cooling system. A water/coolant mix then circulates through the engine, returning to the tank and pipes, where the heat is conducted away at the surfaces in contact with the cooler water. This is one of the most common systems on the inland waterways. This type of system is similar to systems employed in all automotive engines, where a radiator is used instead of a keel tank or pipes.

NOTE: Keel-cooled engines often use an external manifold, which looks identical to a heat exchanger, but without the internal copper 'tube stack', which simply operates as a 'header tank'.

(Direct) Raw-Water-Cooled Systems

(Direct) raw-water-cooled systems will draw water from an external fitting on the hull, through a series of filters and pumps and then through the engine, to exit the vessel via the exhaust pipe or another skin fitting. They may also incorporate a heat-exchanger system as part of the raw-water cooling.

Heat-Exchanger Cooling Systems

Heat-exchanger cooling uses a heat exchanger to cool the engine. One side of the heat exchanger contains the normal engine coolant and the other side has raw water drawn from the surrounding water. The heat exchanger can be incorporated into a water-cooled exhaust manifold or mounted independently. The raw water exits with the exhaust gases after it has been through the heat exchanger.

More information on these systems and identifying the key components are detailed in the chapter on Cooling.

ELECTRICAL SYSTEMS: 12V OR 24V SYSTEMS

Every engine will have an electrical system that delivers power both to and from the batteries and then on to the electrical systems of your vessel. It is important that you are aware of the voltage level of these systems. In most vessels, this will be 12V for the engine and the domestic systems; however, because of the benefits that 24V can offer, many engine electrics are 24V systems with domestics at 12V or 24V. If you do not know your voltage level, there are a number of ways to identify this:

- if there is a single 12V starter battery = 12V electrical system
- if there are two 12V batteries connected in series (*see* Chapter 4) = 24V electrical system.

NOTE: Some larger (usually vintage) engines use two 12V batteries in parallel (*see* Chapter 4) = 12V electrical system. (If you have a bank of 6V or 2V batteries, it is not as easy to identify unless you have the diagram or design information.)

Alternatively, you can check the starter motor and alternator specification – the labels should clearly state 12V or 24V. If all else fails you can use a multimeter to check the voltage level, and take a reading from the starter motor or alternator. (Please ensure the batteries are fully charged and have been left to stand for six hours when taking a reading, as a false reading may cause difficulties.)

GEARBOX AND DRIVE SYSTEMS

Gearboxes come in many shapes and sizes; some are produced by the engine manufacturer, but the majority are from manufacturers of gearboxes for industrial and automotive applications. These manufacturers generally supply across the world and their gearboxes can range from something small enough to power a go-cart to those that power ocean-going super yachts. PRM, ZF (Twin Disc), Hurth, Borg-Warner are some common names, but they can be specific to an engine manufacturer.

It is important to identify the gearbox fitted to your vessel; in most cases this will be a simple case of reading the data plate that is usually present on most. This will contain a type, serial number and ratio and in some cases may also provide guidance on the type of oil needed. Hydraulic gearboxes tend to be larger and have an oil cooler. With classic engines like Lister or an old Ford engine you will find that the gearbox is integral to the engine and swapping for another manufacturer is very difficult without extensive modifications. This can cause issues when problems occur, as most parts are obsolete and having them made is the only option.

Most gearboxes will require automatic transmission fluid (ATF), engine oil or specific gear oil. Sometimes there can be two separate chambers within the gearbox, so it is also worth investigating if there is only one top-up location or two.

COUPLINGS, INCLUDING PYTHON-DRIVES AND AQUADRIVES

Key information about your engine's prop-shaft size and the type of coupling installed can be vital when things go wrong. The chapter on propulsion (*see* Chapter 5) provides photos of these items to assist you in identification. If you do not already know, it is worth finding out what type of coupling is used on your vessel and then to record it on the Engine Details sheet (*see* Appendix). This book will endeavour to highlight the key maintenance tips and advice for each type.

BSS REGULATIONS

The Boat Safety Scheme (BSS) was set up by the Canal & River Trust and the Environment Agency to promote safety and to reduce pollution on the inland waterways. They provide advice on installations and components used on vessels through the identification of hazards, and also promote techniques to avoid risks and prevent boat fires and explosions.

The BSS uses examination to test the systems and appliances on vessels in order to ensure that they meet the minimum safety requirements. The BSS is mandatory on inland waterways craft and is also being rolled out in the Norfolk Broads and is being considered for costal vessels. The BSS certification lasts for four years and is required in order to license your vessel on most navigation waterways.

If you cruise on the inland waterways it is worth ensuring that any work or modifications to systems on your vessel comply with the guidelines. These are available to view online and are continually being updated as new issues are identified.

CONCLUSION

Once you have identified all of the above, take the time to complete and record this information on the Engine Details sheet (*see* Appendix). This will provide you with a quick reference when reading further, as well as being a source of invaluable assistance should you need to obtain technical support when fault-finding. As you work through the book and identify the parts and components on your vessel, add this new information to the Engine Details sheet.

CHAPTER 1

DIESEL FUEL SYSTEMS

INTRODUCTION

The diesel fuel system is one of the most important systems on a diesel engine and requires more

maintenance and upkeep than any other single component or system. It is also the system that is the cause of most breakdowns and can result in expensive repairs when poorly maintained.

Tools required:

- filter wrench
- spanners: 11mm, 13mm, 15mm (selection of others)
- screwdrivers including Phillips
- long-nose pliers or cocktail sticks (for removing seals)
- rags or soak-up mats
- bowl or plastic container.

To assist in understanding and identifying related issues, the diesel system will be broken down into

Filter wrench for removal of filters.

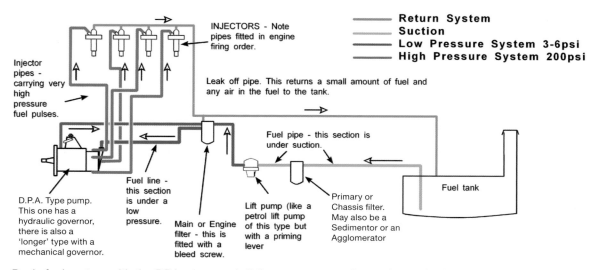

Basic fuel system with the DPA pump and all the components that make up the fuel system.

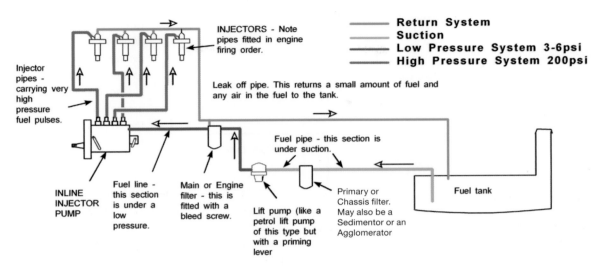

Basic fuel system with an inline injection pump – note that there is no return line from the injection pump.

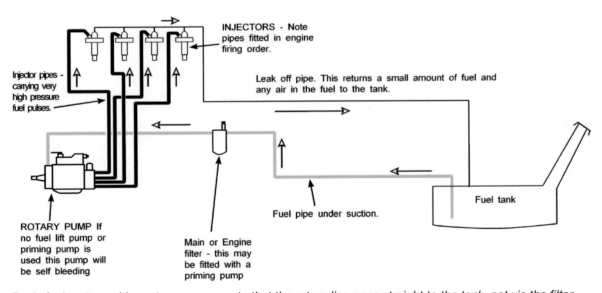

Basic fuel system with a rotary pump – note that the return line goes straight to the tank, not via the filter.

four different systems. These will be referred to as suction fed, low pressure, high pressure and return and are being used for identification of the separate systems only. The fuel system transports fuel from the tank or reservoir through a selection of filters and pumps to deliver clean, pressurized fuel to the engine. However, the type of injection pump will dictate the return system and the components

incorporated within it. The diagram shows the three most common arrangements.

SUCTION-FED FUEL SYSTEM

For our purposes, the suction-fed system covers the components from the tank to the lift pump and includes any pre-filter elements. (It is worth noting

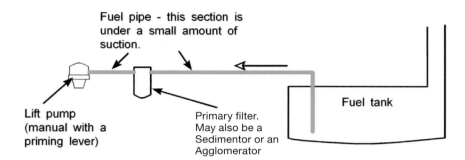

The Suction Fed System

Fuel pipe - this section is under a small amount of suction.

Lift pump (manual with a priming lever)

Primary filter. May also be a Sedimentor or an Agglomerator

Fuel tank

Suction-fed system, showing the fuel system from the tank to the lift pump. This system is under a small amount of suction from the lift pump, meaning that there is little or no pressure pushing the fuel through the system.

that a gravity-fed system would not incorporate a lift pump and would use the head of fuel to gravity-feed the system.) The following section looks at all of the components within this subsystem and describes the type of issues they experience.

This system is under a small amount of suction from the lift pump, meaning that there is little or no pressure pushing the fuel through the system. Therefore, this system is susceptible to drawing air in from any leaking unions or pipes, as there is no pressure to force the fuel out and the slight suction from the lift pump contributes to this process. Air pulled into the suction-fed system will eventually lead to air being transported and trapped in the high-pressure system, resulting in an airlock and causing erratic running or the engine to cut out. It is therefore important to ensure that all unions are

tight and fuel pipes are sound. (So-called self-bleeding systems will cope with small quantities of air in the system, but will eventually succumb to the issue as it gets worse.)

Fuel Tank

The low-pressure system starts at the fuel tank. Its construction and maintenance are important elements of maintaining the fuel system on a boat. Unfortunately, few tanks have all the components shown in the diagram and this can and does result in expensive breakdowns, but aftermarket refits or modifications are not easily implemented. Understanding the impact that tank maintenance can have on the rest of the fuel system is therefore vitally important, and correctly maintaining or managing the fuel in your tank can result in the fuel system running without problems.

Fuel tanks in boats are usually constructed so that the fuel supply pipe for the fuel system is located approximately 25–50mm above the bottom of the tank. The reason for this construction is twofold:

- Most tanks are metal and therefore over time fuel/water/air results in corrosion and rust builds up; this debris drops to the bottom of the tank and remains there.
- The water that is present in the tank, due to condensation and water ingress, naturally separates out and drops to the bottom of the tank when left to stand for periods of time.

TIP: IDENTIFYING FUEL PROBLEMS

Fuel leaks usually appear in the suction-fed system when the engine is not operating, because when the engine is running air is drawn into the fuel system and fuel cannot escape. These leaks are usually very small, so can be difficult to locate.

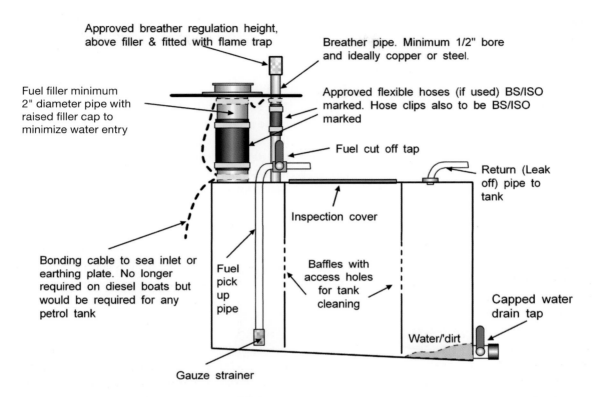

Approved breather regulation height, above filler & fitted with flame trap

Breather pipe. Minimum 1/2" bore and ideally copper or steel.

Fuel filler minimum 2" diameter pipe with raised filler cap to minimize water entry

Approved flexible hoses (if used) BS/ISO marked. Hose clips also to be BS/ISO marked

Fuel cut off tap

Return (Leak off) pipe to tank

Inspection cover

Bonding cable to sea inlet or earthing plate. No longer required on diesel boats but would be required for any petrol tank

Fuel pick up pipe

Baffles with access holes for tank cleaning

Capped water drain tap

Water/'dirt

Gauze strainer

Typical example of a marine fuel tank, including all the elements to allow correct maintenance and cleaning and to prevent some of the common issues that are found with some tank designs.

In both these cases, the outlet pipe position ensures that neither water nor debris is passed through the fuel system, providing a level of protection from these types of issues.

The boat's movement can also cause fuel and debris to mix, so baffles are used to prevent the fuel surging about. The tank may have a sloped bottom (or the whole tank may be sloped). This is to ensure that the water and dirt fall to the lowest part, where they can be drained off using the drain tap (if one is fitted). When the tank has not been designed in this way, the debris will sit along the length of the tank floor.

If a drain tap has been fitted, it must also have a cap to comply with the Boat Safety Scheme. The drain cap is there to prevent filling the bilges with diesel should the tap fail, or be inadvertently opened.

Low-Pressure Pipes

The pipework from the tank to the filters is the start of the suction-fed system. As a general rule, all fuel pipes should be of steel or copper and any joints should be compression fittings. All engines vibrate, no matter how well mounted, so inevitably fuel pipes are subject to the same vibrations. Copper pipes have a tendency to harden over time due to vibration or movement, resulting in them shearing or cracking.

There are a number of arguments that advocate the use of 'rubber' flexible hoses, as these naturally dampen vibrations and also start to show signs of wear long before they actually fail. Using flexible hoses running between the engine and hull pipework can offer a number of benefits, as long as the pipe is compliant with BSS standards.

BSS standards recommend that a hose marked

with BS EN ISO 7840 is used, although hoses marked with SAE J 1257, DIN 4798, or marked with the type of fuel in use are also acceptable. These recommendations are updated regularly and it is always worth visiting the website to check for changes.

Pre-Filters

Depending upon the engine type and the boat

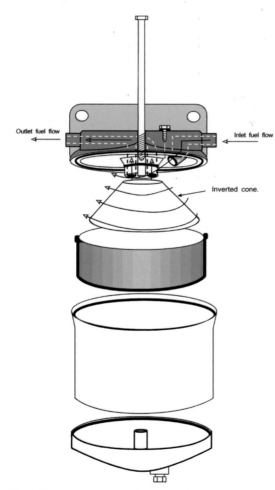

Fuel-filter sedimentor. As the fuel enters, the inverted cone and the angle cause it to swirl around, gaining speed as it moves down the cone. Any large pieces of dirt or water are spun 'centrifugally' to the edge, where they drop into the bowl to await manual removal.

builder, the use of pre-fuel filtering systems varies from one boat to another. It is important to identify which components are installed on the vessel and then to modify servicing requirements to fit.

The sedimentor   This is usually the primary filter in the system (that is, the first after the fuel leaves the tank). Its centrifugal action causes any water and debris to travel to the edges, where they fall into the bowl below. The bowl at the base must comply with BSS regulations. In the past, glass was not allowed, but now plastic is not while glass or metal is, so it is always worth checking the current regulations. New systems like the Fuel-Guard, which is compliant with BSS regulations, provide innovative solutions.

Sedimentor with aluminium body; the butterfly screw on the bottom is used to drain the water.

The agglomerator   This is normally mounted close to the sedimentor, or may be used in place of the sedimentor. The filter housing contains a paper filter made up of a roll of folded and glued paper, so that it forms a series of pockets, open at the top where the fuel enters. The filter is designed to separate out the water that is locked into the fuel. Once separated, the slots cause the fuel and water to spin. The fuel, being lighter, quickly alters its path to flow up through the centre of the element and the water and dirt drain into the bowl.

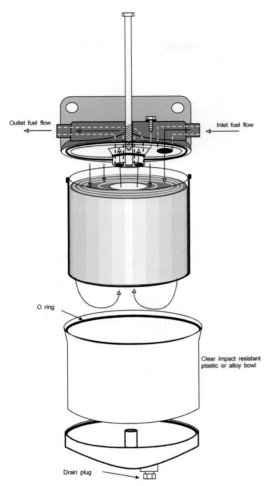

Outlet fuel flow

Inlet fuel flow

O ring

Clear impact resistant
plastic or alloy bowl

Drain plug

Fuel-filter agglomerator. As the fuel travels through the pocket, the water in the fuel is too large to pass easily through the pores in the pockets. The water builds up and gradually increases in size to become droplets. The water droplets are carried downwards by the fuel, which is continually flowing through the filter, then exit through the slots at the bottom of the element.

LOW-PRESSURE SYSTEM

The low-pressure system extends from the outlet of the lift pump up to the injection pump, including all filters and pumps in this section. It is named to reflect the fuel's condition, that is, it is under a small amount of pressure, which increases as the engine

The Low Pressure System

Return

Fuel line - this section is under a low pressure.

D.P.A. Type pump. This one has a hydraulic governor, there is also a 'longer' type with a mechanical governor.

Main or Engine filter - this is fitted with a bleed screw.

Lift pump (some with a priming lever)

Low-pressure system, showing the fuel route from the lift pump through the injection pump and return to the filter. It is under a small amount of pressure, which increases as the engine revs are increased.

revs are increased, to provide an abundance of fuel to the injection pump. The following section looks at all components within this subsystem.

Although this system is still susceptible to drawing in air, the engine will more than likely continue to run even when there is an issue with the fuel system. Where a fractured pipe will draw in air in the suction-fed section, the pressurized fuel is forced out instead and no air will enter this system, so if the bilge is full of fuel the low-pressure system is the most likely place to look for the fault. A fault on this system will still affect the engine running as the fuel pressure drops, causing a lack of fuel supplied to the injection pump.

Engine Fuel Filters and Housings

Fuel-filter housings on the engine come in many different shapes and sizes depending upon the age of the engine. These are fitted directly to the engine and will usually consist of either a spin-on or CAV type of filter. Some of the older filter housings are shown in the diagram. The newer ones simply have a head and seat instead of the whole body.

Fuel Filters

Most engine fuel filters are paper-based elements. The fuel passes sideways through a paper filter that is designed to remove the smallest dirt particles. The sedimentor and agglomerator provide addi-

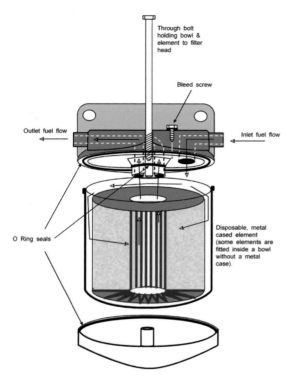

CAV filter housing. The paper element collects contaminants. Water will block the filter, so it is important to change the filter regularly and to clean the bowl.

Older types of fuel-filter housings and sedimentors, some with water traps.

CAV filter housing with priming pump on the housing, usually found on modern engines.

tional pre-filtering to prevent this fine filter becoming clogged too quickly. However, in many cases this is the only filter that is found on the fuel system. The photographs show typical filters used on diesel engines.

A fuel-lift pump filter is often included to prevent dirt entering the pump and causing malfunction. It may be inside the pump as a gauze mesh that can be cleaned, or as a paper-filter element that has to be changed. It can also be external to the pump on the inlet side and vary from full-sized filters to something a little larger than a pipe union containing a gauze filter. This filter is generally overlooked, as many owners and engineers are simply not aware that it exists, which can result in breakdowns being misdiagnosed. Typical engines that use these as standard are Vetus, Barrus and Isuzu, where the

Fuel filters come in all shapes and sizes – this is a typical CAV filter.

Spin-on fuel-filter variations.

Inline fuel filter, which simply connects on to the fuel line. These are usually small components and secondary to the main filter.

Fuel-pump filter – this small filter fits inside the electric fuel pump.

When the pump is mounted vertically (so that the cap is horizontal), this can be the easiest place to check for water when no pre-filtering is present. The electric pump is a sealed for life unit and providing that the wiring is kept in good condition and it has a secure 12V electrical supply, it will run for many years without problems.

filters are inside the electric fuel pumps, or BMC, Perkins and Shanks, where the filter is housed inside the mechanical lift pump.

Fuel Pump (Lift Pump)

Most engines will have some form of primary fuel pump that is used to pump the fuel through the low-pressure system. These come in many different forms, but will either be mechanical or electrical and where a mechanical one fails (especially on older engines where spares may not be available), a replacement electric pump can be easily installed.

The fuel pump is likely to have its inlet and outlet in the main body. It will have a cap fitted with a single bolt or screw. Under this cap is where the gauze strainer and sediment trap, or paper element filter, will be located.

Lift pump (manual) with side priming lever; note the differences in priming-arm location and inlet and outlet connections, and the depth and shape of the cam arm (which sits inside the engine).

This example has an extension bar to help with access.

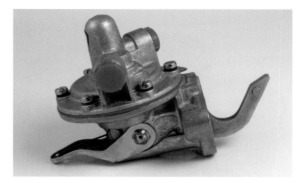

In this example the priming arm is attached at the back.

Electric pump (Facet) 12V, showing fuel-pipe inlet and outlet and electrical connections.

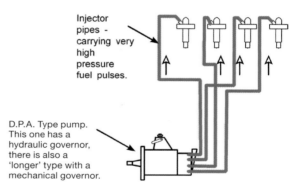

The High Pressure System

Injector pipes - carrying very high pressure fuel pulses.

D.P.A. Type pump. This one has a hydraulic governor, there is also a 'longer' type with a mechanical governor.

High-pressure system, showing the fuel route from the injection pump to the injectors. The fuel must be pumped at extremely high pressures in order for it to atomize when released into the combustion chambers.

HIGH-PRESSURE FUEL SYSTEM

The high-pressure system covers from the injection pump to the injectors and the following section looks at all components within this subsystem. In the high-pressure system, the fuel must be pumped at extremely high pressures in order for it to atomize when released into the combustion chambers. It is very important that the pressure stays consistent. Any leak, as slight as it may seem, can cause a very noticeable misfire.

TIP: SYMPTOMS OF AIR IN THE FUEL SYSTEM

If air is present in this system, it will not pressurize like fuel; instead, air squeezes and compacts, which results in the engine cutting out and shutting down. Symptoms such as knocking, racing and surging are common faults associated with the high-pressure system losing fuel pressure, although the faults can originate from either of the previous systems.

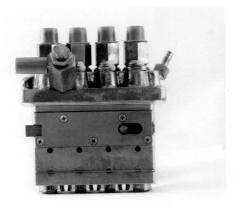

Zexel-type injection pump. All injection pumps perform the same operation, but work in different ways. Typical on Kabota-based engines and newer marine engines.

CAV-type pump, typical on older engines like BMCs and Perkins engines.

Typical injection pump for a Ford engine, usually referred to as a Simms pump.

Injection Pump

The injection pump is part of the high-pressure system and delivers high-pressure fuel to the injectors. The main reason for failure of this element is usually due to fuel contamination, internal rusting caused by water in the fuel, or 'racks' sticking and seals failing from lack of lubrication. Diagnosing issues and learning how to bleed the pump is essential and will be covered later in the chapter. However, where removal and repair are required, this is best left to the specialist as the pump is integral to the engine timing.

Bryce injection pump, typically used on Listers, Petter engines and older engines. There is usually one per cylinder and they are internal to the engine.

Injector types. There are hundreds of different types of injectors and although they all do the same job, they contain different components.

Injectors

Whatever their shape, all injectors work on the same principle. The only things that alter are the nozzle type and how the injection pressure is adjusted. The injectors atomize the fuel and supply it to the combustion chamber.

Occasionally, air can get trapped between the injection pump and the injectors. It can be bled from the system by loosening the high-pressure pipe at the injector end, but this should be undertaken with caution. Should the injectors need overhauling (due to fuel, water contamination or age), they should always be sent to a specialist.

High-Pressure 'Injector' Pipes

The injection pipes are usually moulded to the specific engine requirements and are bought specifically for the engine. As they are under high pressure and connect directly to the engine, they are subject to the engine vibration and can suffer from shearing and cracking. If this happens, there will be a distinct misfire and a noticeable leak, and the engine may start to smoke either blue or black smoke in varying quantities.

TIP: ADVICE ON HOW TO REMOVE INJECTOR PIPES

When removing the injector pipes, it is important to loosen the high-pressure pipes at the pump end first; this provides sufficient movement for the pipe to be moved clear at the injector end. This avoids bending, which would lead to premature failure and trouble when trying to reconnect the pipework. Some engines have rubber bushes fitted to the injector pipes, which are used to dampen vibration, so they must always be refitted, otherwise the pipe is likely to fail through vibration.

TIP: BUBBLING AT THE INJECTORS

Fuel bubbling where the injector enters the cylinder head can indicate that the copper washers on which the injectors are seated are leaking. This is usually due to age and indicates that they need replacing, although they may simply need to be tightened, so this is worth a try first.

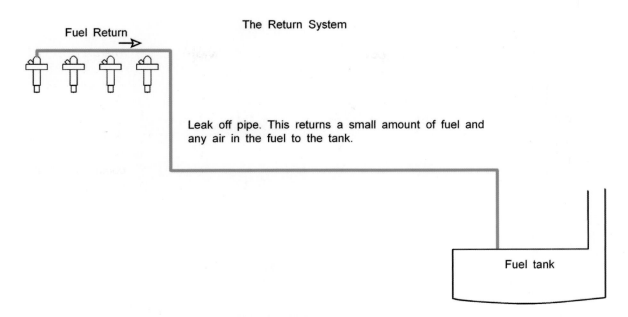

The Return System

Fuel Return →

Leak off pipe. This returns a small amount of fuel and any air in the fuel to the tank.

Fuel tank

Return system returns excess fuel and fuel used to cool and lubricate the internal parts of the injection pump and injectors to the fuel tank (this example shows the return from the injectors to the tank, but there can also be a return from the pump to the filter, or a combination of both).

RETURN SYSTEM

The return system is an essential part of the fuel system. It returns excess fuel and fuel used to cool and lubricate the internal parts of the injection pump and injectors to the fuel tank. The return can come from the injectors to the tank, from the injectors to the engine filter, or from the injection pump to a filter or direct to the tank (*see* diagram 'Low-pressure system'). The return pipe connection on the tank is normally at the top and usually contains a non-return valve (which is also known as a one-way valve).

If the return pipe fractures, the engine running will not be affected, but small amounts of diesel will leak into the bilges, depending upon the quantity of excess fuel being returned from the injection system.

In rare cases, when the non-return valve fails, the return system can bypass the other fuel systems and can lead to unfiltered fuel back-feeding into the injection pump, which can lead to internal damage. It may also cause the engine to cut out or misfire due to too much fuel pressure (common on rotary type/DPA injection pumps). Furthermore, on self-bleeding systems it will compromise its ability to self-bleed, causing issues during routine maintenance. This condition can be identified by turning off the fuel stopcock and loosening the return pipe on the injection pump, to see if fuel leaks out.

HOW TO MAINTAIN THE FUEL SYSTEM

Diesel Bug and Water Contamination

All diesels contain some element of water. Water content in diesel makes the diesel murky or cloudy and can be the first indication of an issue developing.

However, when diesel is left to stand this water separates naturally (around 800ppm 0.08 per cent) and as it is heavier than diesel it eventually drops to the bottom of the tank, and hence why the drain tap on the tank offers real maintenance benefits. Leaving water in your fuel tank will create the ideal conditions for diesel bug contamination to develop and over time will cause internal rusting of the tank.

Diesel sample with water contamination. Note the cloudiness of the fuel sample.

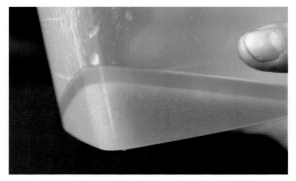

Diesel sample showing emulsified fuel. This is where too much water has become locked in the diesel, causing it to emulsify under pressure.

Diesel sample with no water contamination. Note how clear and bright the fuel sample is.

A severe case of diesel bug. This is the inside of a fuel-filter housing and shows how diesel bug can clog and block the fuel system.

Diesel bug is the term given to the enzymes and bacteria that live off the water in diesel and affect the diesel's properties. There are over 148 different types identified so far. The first noticeable sign of fuel degradation is a fine black dust that is often described as soot, a very strong smell of varnish coming from the fuel tank and the fuel turning darker. There are other lesser known variations of diesel bug, which can show as yellow/orange or pale debris floating in the diesel.

Diesel contamination. This sample was removed from the fuel tank and shows water contamination and the jellification caused by a type of diesel bug.

Visible sludge is an indicator of severe or high-risk contamination and should be treated as soon as possible. Extreme cases can see this develop into a thick black sludge that quickly clogs up the fuel system and stops the engine operating.

Biodiesel and Low Sulphur Fuels (EU Directive)

The introduction of new rules relating to the supply of diesel for recreational boats has stipulated that from 2011 it is an offence to sell diesel that contains more than 10mg of sulphur per kilogram of diesel. This effectively makes the fuel virtually 'sulphur free'. Diesel that is low in sulphur can cause issues with rubber and plastic components in the fuel system because it contains fewer lubricants and eventually this lack of lubrication can lead to faster degradation of these components, resulting in fuel pump and seal failures.

To accomplish the low sulphur content, many suppliers have opted to incorporate biodiesel (when biodiesel is blended, fatty acid methyl esters are added to the mineral diesel). Including biodiesel in the fuel mix can resolve the lubrication issue caused by low sulphur, but it also promotes fuel oxidation and instability, and is incompatible with engine seals and sealants. Its biggest downside is its absorption/attraction of water and formation of emulsions, resulting in higher cases of diesel bug and fuel emulsification.

Fuel Treatments

If there are concerns about diesel contamination, there are many products available to help combat these problems. Most fall into two categories: enzyme-based products that 'lock the water into the fuel'; and chemical-based products that 'drop the water out of the fuel'. The arguments will always remain among manufacturers as to which treatment is best and in the hope of helping users to identify the best products on the market a number of independent tests have been completed, although the results could be concluded as being more confusing than helpful!

Contamination is responsible for hundreds of breakdowns each year and in my opinion enzyme-based products are effective only when there is a small amount of water content, but too much and you run the risk of not removing it, or overdosing and locking too much water into the diesel. Therefore for me a product that provides instant results and stops these problems from recurring is more important, which is why I favour biocide treatments.

TANK MAINTENANCE

Regularly check the fuel-tank filler cap seal and replace if worn, cracked or damaged. Failure to do so will result in water entering the tank through the threads during heavy or prolonged rainfall. As the cap sits lower than the deck, it is important to wipe over to remove excess water before opening the cap to refuel if it has been raining. Any water in the fuel or entering the tank will naturally separate when left to stand and will fall to the bottom of a tank with the diesel remaining above it. If your tank has a drain tap, regularly draining off any water will avoid contamination and the development of diesel bug.

TIP: GETTING A SAMPLE FROM THE FUEL TANK

If there is no drain tap on your tank, drop a clear plastic hose through the filler to the bottom of the tank and seal it with a hand or thumb, then withdraw; this provides a sample of what is the tank. As diesel sits on top of water, you will be able to see clearly how much water is present. If more than 50mm of water is present, use a simple siphon pump (or oil-extractor or electric pump) to extract it. Siphon off the water by pushing the pipe down to the bottom of the tank, extracting fluid until the diesel comes through. Alternatively, many marinas now offer a 'fuel polishing' service to clean the fuel and remove debris and water.

Fuel cap and the rubber seal, which should be checked and replaced periodically.

NOTE: The fuel filler cap seal can be replaced with a rubber O-ring, which you can source from any automotive parts retailer.

The biggest causes of water in diesel are condensation and natural separation, and there are many arguments about how to combat these if a vessel is to be left standing for a period of time. There are really only two choices: if the tank is left empty, you must ensure that you remove any water when returning to the vessel or refuelling; alternatively, the tank can be left full and treated, but you must make sure that you siphon off any water in the bottom of the tank before using the boat. This advice will also combat both diesel bug and the bio-diesel issues associated with water contamination, removing almost all risks without the need to treat the fuel.

FUEL-FILTER CHANGE

Changing a filter is a simple and essential part of maintaining your diesel engine and fuel system. As a general guideline, filters should be changed every 200–250 hours, or every year, even if the boat is used infrequently, unless stated differently in the owner's manual.

Ensure that you have a small container, rags or soak-up mats, the tools you require and the filter at hand before starting. Before beginning, ensure that the fuel stop tap is in the OFF position (usually when the lever is at a 90-degree angle to the pipes). Always start with the filter closest to the tank, then move on to the next one. Undo each filter as detailed below, change the element and reassemble prior to removing the next filter.

Fuel-Filter Service Procedure

These instructions are for general fuel-filter servicing. Variations of filter type follow this section.

- Locate and turn off the fuel at the tap, prior to commencing any filter removal.
- Locate the first filter and identify the type.
- Before removal of the fuel filter, wipe down the filter body/housing so that debris and dirt are not introduced when replacing the filter and make sure you place a rag or oil soak-up mat to catch any stray fuel that may spill in the process. Remove the filter (*see* details below of the different types of filters). If the filter is stuck, follow the instructions in Chapter 2, 'Engine Lubrication'.
- After removing the filter, check the filter head for remains of old rubber seals, especially in the recess of CAV-type heads. Use non-metallic tools such as cocktail sticks to avoid damaging the sealing surfaces. Remove any other seals from the housing.
- Pour the diesel into a small container and inspect for signs of contamination, or place paper filters in the container for inspection. You must ensure that the diesel and any rags or soak-up mats are disposed of responsibly.
- Fit a new filter or replace the filter element, ensuring that all seals are replaced and fitted correctly. Before fitting a seal, always use a little fuel to lubricate (*see* instruction below).
- When all the filters have been changed, turn the fuel supply back on and bleed the system (*see* later in this section for guidance).
- Once happy that all the air is out of the system, wipe down the engine and check that there are no leaks before starting the engine.

- Try to start the engine; it should start as normal. If it fails to start or stops straight away, you need check the system for leaks on the filters, then re-bleed.

NOTE: If the engine does not start, do not crank the engine beyond the time it normally takes to start. This can overheat the fuel/shutdown solenoid on a number of modern engines. The exception to this is the Ford 1.6/1.8 XLD engine, which can only be bled by cranking.

The following provides more guidance on the different fuel-filter elements, how to dismantle and refit, and any special instructions or guidance needed. Locate the one relevant to your installation and follow the instructions. Please see later in this section for photos of typical filters and locations on common engines to assist in identifying the type of filter fitted to your vessel.

Cartridge (CAV) Filter Replacement

CAV filters are removed by undoing the central bolt and changing the cartridge. When removing a cartridge filter, fuel will begin to leak as soon as the centre bolt is loosened. It is therefore essential to have a soak-up mat in position and a container available in which to place the filter and catch the fuel that will be inside the filter – *before* commenc-

ing! CAV filters can have up to five seals included in the filter box and these should be changed every time the filter elements are replaced.

1. Locate the fuel-filter housing and identify the centre bolt that holds the housing together.

2. Locate the main bolt in the centre of the housing. Using a ring spanner, loosen the bolt. Place a container under the filter to catch the fuel.

3. As the bolt is loosened, the fuel will begin to drain into the container.

Seals supplied with a CAV filter. All of these seals have to be located, removed and replaced.

4. Once the fuel filter has drained, undo the bolt fully, whilst holding the bottom of the filter housing. Once undone, the filter body should come away; if it does not, give it a slight twist to assist it.

5. Tip the remaining fuel into the container and inspect, looking for debris and contamination. Clean out the body, by wiping it with a rag or cloth.

6. Fuel-filter housing top, showing the main bolt hole without the bolt.

7. Place the filter into the waiting container and pour any fuel from the fuel-filter housing bowl into the container.

For each seal, hook the old one out with long-nose pliers, or a cocktail stick if it does not come away whole – make sure nothing is left.

8. Remove the seals located in a recess under the top outside edge of the fuel-filter housing (the top housing in the photo has been removed from the engine in order to show where the seal fits).

9. Remove the seals from the bottom section of the housing on the outside edge of the bowl.

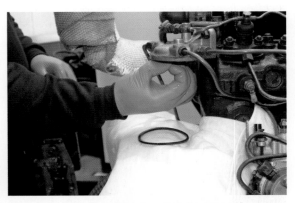

10. Check the fuel-filter housing head and remove the small seal for the bleed bolt in the top of the housing and the seal for the main bolt.

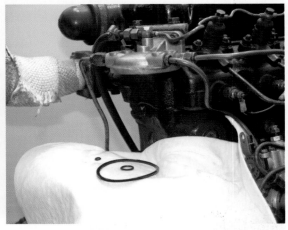

11. This shows the seals removed from the fuel-filter housing head and the seal that sits on the main bolt.

12. *Fit the new main seal into the fuel-filter housing head.*

14. *This shows the seals in place and seated correctly.*

Wipe the area clean and, before fitting a new seal, use your finger to rub a light covering of fuel around the rubber. This helps to stop it sticking and creates a better seal when the filter is fitted. Finally, push the new seals into place, ensuring that they are not twisted.

13. *Fit the centre seal into the fuel-filter housing head.*

15. *Fit a new seal to the fuel-filter housing bowl and then slot the new filter on to the bowl.*

16. *Match the fuel-filter housing bowl and filter up to the fuel-filter housing head. Whilst tightening the centre bolt, twist the body; if it scrapes it is rubbing against the metal head unit and is probably out of line. In this case, undo the centre bolt and realign the body so that you can feel it seating and twisting against the rubber seal.*

17. Once secure, tighten up the main bolt with a ring spanner. Take care not to overtighten the centre bolt, because they can easily distort and will cause fuel to leak around the bolt.

Once the filter has been changed, you will need to bleed the fuel system (*see* later in the section for guidance), then run the engine for a short time whilst feeling around the filter for leaks. Once it has run for up to half an hour, turn the engine off and check for leaks once more.

Spin-On Filters

Remove the filter from the engine by twisting it anti-clockwise, using a filter wrench initially and then unscrewing it by hand. Fuel will immediately begin to escape and run down the engine. Place the old filter in a container.

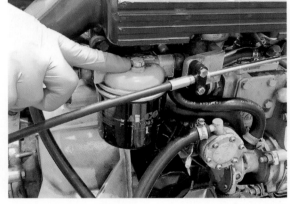

1. To remove a spin-on filter – first locate the fuel filter on your engine.

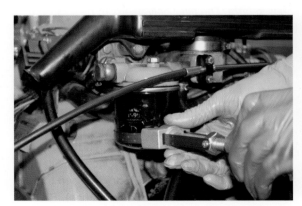

2. Using a filter wrench place the strap around the filter and tighten the strap by winding up the adjuster on the filter wrench.

3. Rotate the filter anti-clockwise to undo it, continue repositioning the wrench as required until the filter is able to be turned by hand.

4. *Remove the filter wrench and unscrew by hand. Have a container handy to catch fuel as the filter is removed.*

5. *Pour the contents of the filter into a container and inspect for particles and contamination. Check that the filter seal has come away with the filter. If not, check filter head and remove.*

6. *Using the new filter, match up to the filter head. Turn the filter until you feel resistance and the filter feels secure. The filter only needs to be hand-tight – you do not need to use any tools.*

Always check the engine filter 'seat' to make sure that there is nothing stuck that might obstruct the seal when fitting the new filter. If a whole or partial seal is found, remove it and wipe the area clean. The new fuel filter will be supplied with a new seal fitted to it. Always check to ensure that it is whole and has no cracks or breaks before fitting the filter. Smear a small amount of fuel over the seal and screw the filter back into place. The filter should only require hand-tightening. No tool should be required; if you need to use a tool, it is likely that the filter seal is not seated correctly.

Check that the filter has little or no sideways movement as you spin it on, as this can indicate that the filter might have the wrong thread and therefore is incorrect for your engine. Once the filter has been changed, it will be necessary to bleed the fuel system (*see* later in section for guidance). Run the engine for a short time whilst feeling around the filter for leaks. Once it has run up for haf an hour, turn the engine off and check for leaks once more.

Water-Trap Filters

Undo the top of the housing and separate the filter, empty the water and change the cartridge, remembering to change the seals. The only difference between this filter and the CAV type is that the bowl is larger to accommodate the water.

Please note the BSS requirement that any filter which has a plastic drain bolt must to be changed for a metal one.

Agglomerator with filter and water trap; the butterfly screw on the bottom is used to drain the water.

TIP: CHECKS TO BE MADE WHEN CHANGING FILTERS

Before fitting a new filter, always check that the seal, or part of it, is not stuck on the filter head, as this will result in a leaking filter, or air being pulled into the fuel system. Always pour the contents of the fuel filter into a container to see if there is any evidence of diesel bug or contamination. Visually check the paper elements.

BLEEDING SEQUENCE: TRADITIONAL

The fuel system will require bleeding whenever filters or other components are replaced, or when the fuel tank has been run low. Most modern engines will self-bleed: for engines with electric lift pumps simply turn the ignition key to the run position for 20sec (first key position); for engines with manual priming lever/plunger operate until consistent firm pressure is felt.

NOTE: There is always 25–50mm of fuel left in the tank when empty.

Engines with mechanical lift pumps without manual means of priming will need to be cranked over (turn the key and attempt to start); do not crank for longer than it would normally take to start. Otherwise, follow the manufacturer's recommendations. Note that this should be undertaken with caution, as it can quickly drain the batteries. To prevent unwanted starting during bleeding (ideally you do not want to start the engine until all the air is bled from system), either disconnect the electrical connector on the fuel/shutdown solenoid, or leave the manual stop control pulled out. Reconnect the solenoid or push the stop control back in when bleeding is complete.

Locate the Lift Pump

To find the lift pump, trace the pipework from the fuel tank. It will be the first engine-mounted component in the system and is usually located on the side of the engine (the accompanying photos show the varying locations). Once the lift pump is identified, locate the manual pumping arm. It protrudes from the side or bottom of the pump and will be used to pump fuel manually when bleeding the system (note that an electrical pump does not require bleeding). In some applications, a bar may be fitted to the priming lever to allow you to prime from above the pump.

1. Manual lift pump with arm at rest.

2. Lift pump with arm raised. Lift and raise to pump the fuel manually through the system.

3. An engine lift pump on a Beta engine, with the arm on the side for priming the pump manually.

4. There are different types of manual lift pump. With this one, the arm is attached to a rod that allows the user to prime the pump (pump the fuel) from above.

Locate the Filter Bleed Points

Locate the smallest bolt/screw on the top of each filter, as each filter will need to be bled to enable the fuel to flow to each location. The bleed screws on the filters are the most effective bleed points. On a BMC-type CAV filter, you may find that it does not have a screw to bleed the system, so in this case use the leak-off pipe banjo bolt (the one small screw on the top is the centre 'hold together' bolt – do not use this).

To bleed the fuel system at the fuel filter, locate the bleed bolt and, using a spanner, loosen until it can be unscrewed by hand, but do not remove it.

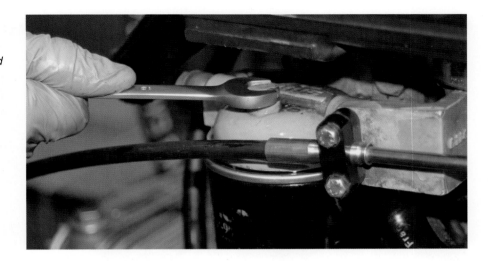

Locate the Union/Bleed Screw on the Injection Pump

Locate the injection pump, then follow the fuel lines from the filters and lift pump. Continue until the main injection pump is reached; this will usually be attached to the engine block. Locate the union on the pipework that enters the injection pump;

alternatively, locate the bleed screw on the injection pump. Be aware that there are a number of screws on the injection pump, so it is important to identify the right one. There are photos of a number of different engines at the end of the section to assist in locating the correct one on your engine.

Location of bleed screw on CAV DPA injection pump.

Location of other bleed screw on the opposite side. One side will be easier to access and bleed than the other.

Locate the Injector pipe nut, loosen (one turn) and allow the air to escape, then tighten and move to the next injector – each one must be bled.

Locate the Union/Nut on the Injector Pipe

Follow the fuel lines from the injection pump up to the injectors. There will be an injector for each cylinder, therefore a 3-cylinder engine will have three injectors and three fuel lines, a 4-cylinder engine will have four injectors and so on. The main nut that connects the injection pipe to the injector should be loosened, then the engine turned over until fuel and air leak out. Once only fuel is present, tighten the nut and move on to the next injector. All injectors will need to be bled before the engine is started.

Bleeding sequence and actions:

- Place a rag or a pot under the filter to catch any diesel fuel that may be spilled. Have some rags ready for cleaning the area and yourself; Latex gloves or barrier cream will prevent skin issues.
- Loosen the bolt using a spanner and then undo by hand, but do not remove the bolt.
- Pump the priming lever on the lift pump (or priming plunger on the filter), whilst watching the bleed screw on the filter. (Alternatively, turn the ignition on if the engine has an electric lift pump.)
- Air will hiss out of the bleed screw and it may blow bubbles. When pure fuel with no air bubbles

is coming out of the bleed screw, the system has been successfully bled up to this location. Tighten up the bleed screw and move on to the next bleed location.

1. Using the priming lever on the lift pump (or priming plunger on the filter), pump it whilst watching the bleed screw on the filter. Alternatively, turn the ignition on if the engine has an electric lift pump to force the fuel through the system.

2. *Air will hiss out of the bleed screw and it may blow bubbles. When pure fuel with no air bubbles is coming out of the bleed screw, the system has been successfully bled up to this location. Tighten up the bleed screw and move on to the next bleed location.*

3. *To bleed the CAV filter, locate the banjo bolt on the top of the CAV filter housing.*

4. *Using a spanner, undo the bolt by turning until it is loose enough to unscrew the bolt by hand, but do not remove fully.*

5. *Using the priming lever on the lift pump (or priming plunger on the filter), pump it whilst watching the bleed screw on the filter. Alternatively, turn the ignition on if the engine has an electric lift pump to force the fuel through the system.*

6. *Air will hiss out of the bleed screw and it may blow bubbles. When pure fuel with no air bubbles is coming out of the bleed screw, the system has been successfully bled up to this location. Tighten up the bleed screw and move on to the next bleed location.*

- Repeat this sequence on any other filters in the system, then move on to the next element.
- Move to the injection pump and loosen the bleed screw if there is one (three turns). If no bleed screw is visible, loosen the union that connects the fuel inlet pipe to the injection pump. Continue pumping the lift pump until the air ceases to be expelled through the union or bleed screw and a clean flow of fuel is present. Then retighten the union or bleed screw. Do not overtighten.

1. Using a spanner (usually 11mm or 13mm) loosen, but do not completely remove, the bleed screw. Three turns is usually sufficient.

2. Unscrew by hand when loose enough, but do not remove. Pump the priming lever on the lift pump (or priming plunger on the filter) whilst watching the bleed screw. Alternatively, turn on the ignition if the engine has an electric lift pump.

3. Air will hiss out of the bleed screw and it may blow bubbles. When pure fuel with no air bubbles is coming out of the bleed screw, the system has been successfully bled up to this location. Tighten up the bleed screw and move on to the next bleed location.

If you are bleeding the engine after it has cut out, or are servicing the fuel system, you will also need to bleed the high-pressure system:

- Locate the injector pipes. Follow the pipes from the injection pump until they enter the engine.
- One at a time, loosen the injector pipes at the injector end, that is, where the pipe connects to the injectors and enters the engine (one turn is sufficient).
- Crank the engine over until fuel starts to spit or drip from the injector union. This is not a bleeding point, but allows any air still left in the system to be forced out.
- Once fuel is spitting from the injector union, retighten the injector pipe and move to the next one.
- The engine should now start. Always preheat the engine for 10–15sec before starting.

NOTE: After bleeding, the engine may take a few seconds longer to fire as the system will need time to repressurize. On old engines (BMC, Lister and so on) if the engine does not fire, stop cranking and begin the bleeding process again.

TIP: BLEEDING A HIGH-PRESSURE SYSTEM

The only way to bleed a high-pressure system is to turn over the engine, which requires plenty of power. It is always best to avoid turning over the engine when bleeding the previous systems, in order to conserve power, and is especially important if the system needs to be re-bled.

SIMPLE BLEEDING: INLET PIPE

On modern engines (or as confidence is gained) some of the above steps can be missed and the fuel system can be bled at the bleed screw, or from the inlet pipe on the injection pump. Modern engines are 'self-bleeding', so simply locate the union or bleed screw on the injection pump, turn over the engine and the system will purge the air. Once fuel is flowing freely, retighten and attempt to start the engine. The following six photographs show various types of bleed points.

Vetus (Mitsubishi), showing the location of the bleed bolt on the filter housing.

Location of the bleed bolt on the injection pump (Vetus).

Location of the bleed bolts on the injectors (Vetus).

Showing the lift pump being primed whist the filter is bled.

Location of the bleed bolt on the filter (Beta).

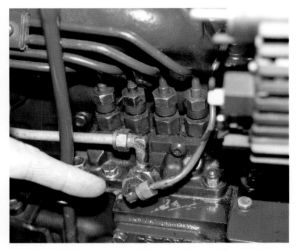

Beta (Kubota), showing the location of the bleed point, which is on the injection pump return, as the injection pump bleed point is not accessible.

COMMON PROBLEMS AND SOLUTIONS

Following are some of the common problems and their solutions:

- The exhaust smokes whilst the engine is turning over, but does not start. This indicates that there is fuel present – increase the preheat in order to fire the engine. (The engine will fire with a preheat of 20–30sec, even on a fifty/fifty mixture of water and fuel.)

- There is no smoke, or only wisps of smoke appearing, and the engine turns over but does not start – there is no fuel! Check that the stop cable has not been pulled out, then re-bleed the system. This can happen when the injector pump has not been bled enough, as it can take a while to completely expel all the air.
- Fuel leaking from a filter – check that the old seal has been removed and the new seal is seated correctly and is not twisted.

CHAPTER 2

ENGINE LUBRICATION SYSTEM

INTRODUCTION

The lubrication system is one of the most undervalued and dismissed elements of the engine, but it is one of the most important in terms of operational performance and engine life. Poor maintenance of the lubrication system, or failure to note a critical change, can result in the most expensive repair costs and in some cases a complete engine failure and need for replacement.

The lubrication system is relatively simple, with only a few components that need to be maintained. However, oil selection and correct filter identification are key to ensuring the longevity of the engine and maximizing its capabilities. The following chapter provides a description of the main components and their interactions and how to undertake a basic service of the oil system. Before embarking on a service, it is advisable to invest in good quality soak-up mats, absorbency kits and spill socks. This will help to keep your engine room clean, reduce the risk of an environmental spill and make working on your engine a lot easier.

Tools required:

- filter wrench
- oil suction pump (siphon pump)
- selection of spanners to suit the engine
- container for disposal of old oil
- rags or oil soak-up mats.

BASIC OIL LUBRICATION SYSTEM

The lubrication system's main function is to deliver oil at the right pressure and temperature around the engine and to ensure that the internal components are operating smoothly. The benefit of this lubrication is to reduce friction of the moving parts, which in turn results in better fuel economy and component wear.

The pump is the heart of the lubrication system and pumps oil around the engine. Oil is drawn from the sump into the oil pump, then travels through a filter (the oil is not filtered before the oil pump). This filters out any dangerous particles in the oil before being transported to critical components through pipes and oil galleries to remote locations like the pistons and bearings. Every element in the engine needs lubrication to keep it operating correctly and to prevent wear and deterioration, as well as to transfer the heat created by friction between metal components.

All oil systems will have a pressure switch fitted to monitor engine oil pressure. Once the oil has completed its cycle of lubricating and cleaning the engine, it returns to the sump to begin its journey once more. Should the oil pressure fall to a danger-

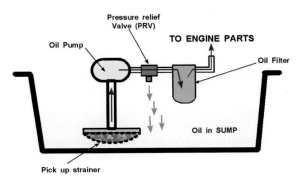

Main components of the engine oil system.

TIP: IDENTIFYING FUEL LEAKING INTO THE OIL SYSTEM

If you find the engine oil level increasing, it is likely that there is diesel leaking into the system. This is usually as a result of a lift pump, or injector pump seal failure. On some engines, for example Lister, Lombardini and Ford, where the fuel system runs internally to the engine, it can be due to a fuel pipe fracture, injector or injection pump failure. The oil will have a 'diesel' smell and if rubbed between the fingers will feel less viscous. The engine should not be run and will need to be flushed and the problem located and resolved as soon as possible to avoid internal engine damage.

and drain the oil through the dipstick. An oil-siphon pump is a very worthwhile investment, as it can also be used to extract water from contaminated fuel, or even help to empty bilges of contaminates.

Oil sump, showing the oil sump nut, which is usually inaccessible on a boat installation.

ous level, the oil-pressure switch will activate a warning lamp or display a warning on a gauge.

It is worth being aware that the oil is only filtered after the oil pump, therefore the dirtier the oil becomes the more the pump will wear and a good percentage of engine failure can be traced to oil pump-related faults. Change the oil regularly and the engine life will increase dramatically.

MAIN COMPONENTS

The engine sump is simply a local oil reservoir and all oil returns to this location once the engine is stopped. If this was in an automotive application, an oil change would be undertaken by removing the sump plug (accessed from under the vehicle) and draining the oil along with any deposits or debris. However, as it is the lowest part of the engine, it is usually inaccessible on a boat and therefore many marine engine manufacturers incorporate a manual oil-extraction pump to overcome this issue. Unfortunately, on many older 'converted' engines, the only way to remove the oil is to use an oil-siphon pump

Oil-sump pump, typically brass and fitted to the side of the engine.

Dipstick location on a Vetus engine.

Dipstick location on a BMC engine.

The dipstick is used to check the level and condition of the oil whilst the engine is shut down. It is important to identify where on your engine this component is located in order to check oil levels. When checking the oil level, always remove the dipstick, wipe clean and then reinsert to get a true reading. Remember that the oil level will vary depending upon how long the engine has been shut down. Checking first thing in the morning, after most of the oil has drained back to the sump overnight, will give the best indication; compare this reading to one taken shortly after the engine has shut down, as this will show a slightly lower oil level. By doing this, you will come to know what is normal for your engine.

The oil filter's primary job is to filter the oil before it passes to the main engine 'arteries'. It removes any fine particles to ensure the smooth lubrication of the moving components in the engine. Many engine manufacturers will supply genuine parts for servicing. However, these can be relatively expensive and there are cheaper options available from

Dipstick location on a Beta engine.

Various oil cartridge filters.

Various oil spin-on filters.

motor factors, but, as a cautionary tale, bear in mind that some filters are not directly comparable. Many marine filters can incorporate a bypass valve and anti-drain valve with different pressure settings than an automotive filter. For this reason, it is always worth checking the technical specifications before using an 'unknown' filter.

The pick-up point for the oil is in the sump and it usually incorporates a strainer, which captures any large debris that has been transported through the engine and has ended up in the sump. Typically, this debris will contain metal filings, swarf, clumps of carbon and old oil. The strainer protects the oil pump by preventing the debris being picked up and transported into the pump.

If the strainer becomes blocked, unfortunately there may be no indication or warning of the issue occurring. However, the oil pressure will drop and though the gauge or alarm may sound, unfortunately it can quickly result in an engine running dry. The only way to clear the blockage is usually to remove the engine in order to gain access to the sump.

If the bypass valve in the filter operates (allowing oil to bypass a blocked filter), you may find that the oil-pressure light comes on when the engine is hot and at idle, but clears when the engine revs are increased. This is because pressure is lost opening the valve; changing the filter will resolve this issue.

The oil pump is usually integrated within the engine and will normally outlive the engine as long

as the oil is regularly changed. However, when there are issues it can result in an engine experiencing sudden oil-pressure drop (this will show on the oil-pressure gauge or lamp) and subsequent engine failure. Any oil-pressure drop needs to be investigated and diagnosed correctly. Although it may prove simply to be a sticking/dirty pressure relief valve, faulty sender, lamp or gauge, it is always advisable to bring in a specialist.

Should the pressure drop be due to a blockage or pump failure, it may be possible to access the pump from the sump, but if not it will require a complete engine strip to replace the pump. In many cases, other components in the engine will also have suffered damage unless the fault was quickly identified. Again, a complete engine strip will be required to assess the damage.

The lubrication system may incorporate an oil cooler, which allows the cooling system to remove excess heat from the oil and ensures that the oil remains at the optimal operational temperature. These are common on turbo engines and air-cooled engines. Oil coolers are commonly installed as part of the cooling system, but can be remote from the engine. Remote types (along with remote filter installations) use armoured high-pressure hoses, which can fail and so should be part of routine checks. Spotting external leaks early could prevent engine damage due to lack of oil and prevent the need to clean out a messy bilge.

TIP: OIL IN THE COOLING SYSTEM

If there is oil present in the cooling system, before assuming that a cylinder head gasket has blown, if an oil cooler is installed remove one of the water outlets. If oil comes out of a water connection, this is a clear indication that the oil cooler has failed and needs replacing.

OIL TYPES AND RATINGS

The first place to look for the engine oil specification is in the engine manual, because boat engines often operate their oil at totally different temperatures than vehicle/industrial engines. This is especially true of inland boats, which may only be drawing 1 or 2 horsepower from their engines. In a car, the oil spends most of its time at well over 100°C, so any condensation boils off. Inland boats often never reach that temperature (this is oil temperature – not water temperature).

Powerboats at sea are usually fitted with an oil temperature gauge to ensure that the oil does not overheat, as the oil is operating at higher temperatures due to the demand for power. You will note when cruising on rivers in a heavy flow that the engine is working harder and at a higher temperature. Smaller engines will overheat quickly and where possible should be recognized as a warning sign that the conditions are not ideal for cruising.

When purchasing oil it is important to look for the Society of Automotive Engineers (SAE) viscosity rating and ensure it is correct for your engine. Oils are separated in to mulitigrade and monograde oils. Multigrade oils satisfy two viscosity ratings and will have a low temperature and a high temperature rating, that is, 15W:40. Most marine engines do not take synthetic or semi-synthetic oil.

When an engine is started, the oil is cold and has a 'thick' consistency, but the oil needs to have a 'thin' viscosity to allow it to lubricate the engine even when cold. When the oil heats up it has a 'thin' consistency, but needs to have a viscosity that continues to lubricate adequately and therefore needs more viscosity. The numbers indicate how the oil will perform in different temperatures, for example the higher the number the thicker the oil:

- SAE 30 – a monograde oil; behaves like a 30-grade oil regardless of temperatures. It is usually recommended for older engines and ones that benefit from a 'thicker' consistency.
- SAE 15W40 – a multigrade oil; when cold it behaves like a 15-grade oil and when hot it behaves like a 40-grade oil. This is the standard oil for most applications, although it is important not to use 'synthetic' oil.

Viscosity, although explained as thickness, actually represents the friction properties of the oil. Therefore it is important to note that the thickness does not change with temperature, just the friction properties. Also, the SAE guide does not guarantee the thickness or quality of the oil and therefore oils of the same SAE may have different consistencies based on the quality.

American Petroleum Institute (API) specifications describe how the oil performs under specific conditions. The letters API will be followed by one or two groups of letters. Groups that start with S are petrol engine specifications, so we can ignore them. The ones that start with C are for diesels. The later in the alphabet, the higher the oil specification, so for new diesel engines CG or CH may be specified. However, in inland use the marinizer may well specify a CC, CD or CE oil. The oil you choose should display both the SAE (viscosity) number and an API spec. Please note that a lower-grade oil like CC does not have the protection properties designed for overheating or high-rpm engines.

SERVICE TIMES AND FREQUENCY

The servicing requirements of an engine do depend on the usage, number of hours run, age of the engine and several other factors. If you have an engine manual, this will detail how often an engine service should be undertaken. If there is no manual present, the general rule of thumb is that the engine oil should be changed annually, irrelevant of how many hours the engine has done. The ideal time for an oil change is at the end of a cruise, or just before you lay up the boat for a period of time. Due to waterways closures and the UK weather, this usually falls in late autumn or early winter and the engine benefits from the anti-corrosion additives that are contained in the 'fresh' oil, providing more protection whilst not in use. Servicing requirements are detailed in the last section of this book.

HOW TO MAINTAIN AND SERVICE

Changing Oil and Filters

To change the oil in the engine, you will need to locate the best way to remove the oil. On some engines, there will be a sump pump (usually brass), but although all engines will have a sump plug, in many cases this will not be accessible due to it being at the lowest point with little space around it. If there is nothing on the engine to assist in removing the oil, procure an oil-extractor pump from a local automotive parts retailer. These come in a variety of sizes and are supplied with various small diam-

eter pipes that are used to draw the oil out via the dipstick hole.

You will need to identify the oil filter and the type of filter fitted. If this is not a new engine, you might note dents present on the old filter, which can be an indication that it has been overtightened in the past. Dents will not affect the filter operation and are therefore nothing to worry about, but can result in the filter being very difficult to remove when it comes time to service the oil system. To remove an overtightened filter, or one that has been in place for a long time, spray with WD40 or similar, leaving it overnight to soak in before attempting to remove. If this is not successful, follow the instructions provided later in this section.

Follow the simple procedure below to undertake an oil service. Instructions on how to remove different types of filters are provided in the section below.

Engine oil service procedure:

- Run the engine for 15–30min to make oil extraction easier. An engine flush can be used during this run-up period to emulsify any heavy deposits (once every three services is a good idea).
- Remove the old oil, using the oil-sump pump or oil-extraction pump. Make sure you have rags or

Syphon pump and pipes in use. Insert the pipe and draw out the oil by pumping the handle.

Use the correct size pipe to insert in to the dipstick hole.

oil soak-up mats to catch any stray oil that may spill in the process. If you are using an extraction pump, have a bucket handy to place the oil pipes and allow them to drain.

- Place the oil in a container and ensure that you dispose of it responsibly.
- Before removing the oil filter, wipe down the filter housing so that debris and dirt are not introduced when replacing it and remember that the oil will be hot and care should be taken. Remove the filter (*see* details below of the different types of filters) and replace the element and seals.
- Replace the filter-changing seals where required; smear a small amount of oil around the seal before refitting to the engine.
- Refill the engine with the correct oil (using a funnel will help) and remember to pour slowly so as to allow the oil to drain into the engine without backing up and spilling.
- Clean the components and check that there are no leaks before starting the engine.

The following provides more guidance on the different oil-filter elements and how to dismantle and refit, plus any special instructions or guidance needed. Locate the one relevant to your installation and follow the instructions. *See* later in this section for photos of typical filters and locations on common engines to assist in identifying the type of filter fitted to your vessel.

Cartridge Filter Replacement

Always run the engine for 15–30min to heat the oil and make it easier to remove and then turn the engine off before commencing servicing. Always remove the old oil before attempting to remove the filter (*see* previous procedure). When removing a cartridge filter, oil will begin to leak as soon as the centre bolt is loosened. It is therefore essential before commencing to have a soak-up mat in position and a container available to place the filter in and catch the oil that will be inside the filter.

The new filter cartridge should come with a selection of seals. Wipe each area clean before fitting the new seals. Select the correct one and rub a light

1. *Locate the oil-filter cartridge on the engine, usually mounted vertically with the filter head attached to the main engine block.*

2. *Locate the nut under the main bowl and unscrew using a spanner and then once loose undo by hand. Undo the centre bolt and the filter body will come away, if it does not, give it a slight twist to assist it.*

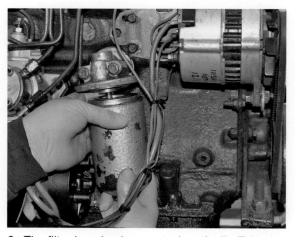

3. *The filter housing is mounted vertically. Take the weight of the bowl when it comes away from the filter head as it will be full of oil.*

4. View of the inside of the filter housing with the oil filter still in position.

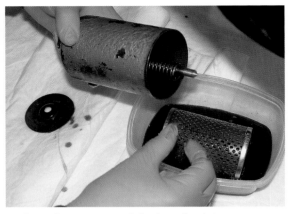

5. Pour the contents of the housing into a container.

6. Place the filter in the container and remove the spring and top hat washer (there is a felt washer on some), which are located on the centre bolt, inside the filter. Clean out the body, by wiping it with a rag or cloth. Check the seal under the head of the centre bolt and replace if damaged.

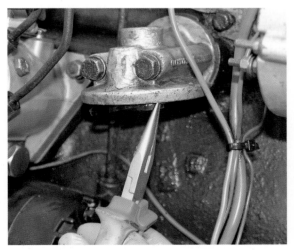

7. Look in the filter head and locate the rubber seal that sits on the inside edge. Use a tool or cocktail stick to remove it.

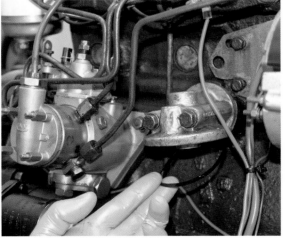

8. Remove the seal and then check the head to make sure it is clear of debris and that the seal has come away whole.

covering of oil around it; this helps to stop it sticking and creates a better seal when the filter is fitted. Finally, push the new seal in its place and check that it is not twisted.

The large washer, which is shaped like a wide-brimmed hat, sits on top of the spring in the same way as a hat sits on your head.

11. Slide the new filter cartridge into the housing with the bolt through the middle. It is now ready to be refitted.

9. Thread the bolt into the housing body and slide the spring on to it, so that it sits in the bottom of the bowl.

12. Locate the large seal from the new filter box and match it up to the filter head.

10. Slide the top-hat washer on to the bolt, making sure that the raised centre is facing you. Slide it down until it sits on top of the spring.

13. Gently push the seal into position around the rim of the filter head.

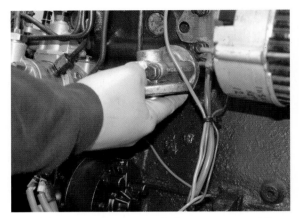

14. *Using a small amount of oil, smear around the seal and also check that the seal is in place and not twisted.*

15. *View of the filter head from below.*

16. *Position the body over the centre bolt and press up to the seal. Whilst tightening the centre bolt twist the body. If it scrapes, it is rubbing against the metal head unit and is probably out of line. In this case undo the centre bolt and realign the body so that you can feel it seating and twisting against the rubber seal.*

17. *Once the filter feels secure and well seated, tighten the bolt with a spanner. Take care not to overtighten the centre bolt, because it can easily distort and will cause an oil leak around the bolt.*

Reassemble the spring and washers on to the centre bolt within the body and fit the new element in the body of the filter. Then fit the new seal to the filter head.

Once the filter has been changed, refill the engine with clean oil by locating the engine oil filler cap and pouring oil in until it shows or is between the two marks on the dipstick (or the mark and the tip of the dipstick). Clean all the old oil from around the engine, then give it a good wipe around before starting.

When you are happy, start the engine, but do not rev it until the oil-pressure warning lamp goes out. Run the engine for a short time, feeling around the filter for leaks. Once it has run up for half an hour, turn the engine off and check for oil leaks once more whilst the engine is warm. Finally, check the oil level on the dipstick (it will have dropped, as some of the oil is now in the filter), but remember that the engine is warm so the level only needs to be between the max and minimum marks.

Spin-on Filter Replacement

Always run the engine for 15–30min to heat the oil and make it easier to remove, then turn the engine off before commencing servicing. Always remove the old oil *before* attempting to remove the filter. It is essential to have a soak-up mat in position and a container to place the filter in and catch the oil *before* removing any components.

Remove the filter from the engine by twisting it anticlockwise, using a filter wrench initially and then unscrewing it by hand. Oil will immediately begin to escape and run down the engine. Place the old filter into a container.

Locate the rubber seal, which should be on the top of the old oil filter, but is sometimes stuck in

3. Use the filter wrench initially and then unscrew by hand. As the filter becomes loose, oil will start to escape.

1. Locate your spin-on oil filter. If mounted horizontally, place a container under it.

4. Remove the filter completely from the filter head and place into a container. Check that the filter seal is not still attached to the filter seat.

2. Using a filter wrench, tighten it on to the filter and twist anti-clockwise to undo.

5. If the seal is still attached, remove it for disposal with the filter.

the filter seat. Always check the engine filter seat to make sure there is nothing stuck that might obstruct the seal when fitting the new filter. If a whole or partial seal is found, remove it and wipe the seating area clean.

The new oil filter will be supplied with a new seal attached; always check to ensure that it is whole and has no cracks or breaks before fitting it. Smear a small amount of clean oil on the filter seal and then screw the filter back into place. The filter should only require hand-tightening, so no tool should be required. If you need to use a tool, it is likely that the filter seal is not seated correctly, or the filter itself is faulty. Check that the filter has little or no sideways movement as you spin it on, as this can indicate that the filter might have the wrong thread and therefore is incorrect for your engine.

NOTE: Prefilling the filter with clean oil before fitting is an option. However, it only works well if the filter hangs vertically when installed. Also, any oil introduced will not be filtered, so any detrimental particles in this oil will travel and might cause wear within the engine.

Once the filter has been changed, refill the engine with clean oil by locating the engine oil filler cap and pouring oil in until it shows at max on the dipstick. Clean all the old oil from around the engine and give it a good wipe.

When you are happy, start the engine, but do not rev it until the oil-pressure warning lamp goes out.

Run the engine for a short time, feeling around the filter for leaks. Once it has run up for half an hour, turn the engine off and check for oil leaks once more whilst the engine is warm. Finally, check the oil level on the dipstick (it will have dropped, as some of the oil is now in the filter), but remember that the engine is warm so the level only needs to be between the max and minimum marks.

Removing a Stuck Filter

It is never advisable to use a screwdriver inserted through the filter body to assist in the removal of a stubborn filter – unfortunately as you try to twist it off, it is likely that it will split open like a sardine can and leave very little alternatives for removal. If the filter is not moving with the use of a strap/chain wrench or other suitable tool, use penetrating oil regularly over the course of a few day, then attach a filter wrench and gently tap with a hammer. It will eventually loosen if this process of penetrating oil application and regular attempts to remove is followed.

Alternatively, place an old screwdriver against the folded rim at right angles to the rim. Using a mallet, or soft hammer, strike the screwdriver a few times to dent the rim. Now angle the screwdriver to the right and continue with firm blows. This will 'jar' the filter, break the contact and with a bit of luck loosen the filter so that you can get it off. Be extremely careful not to scratch or damage the seating face, as this will result in an irreparable oil leak.

Fit the new filter (checking that the seal is in place), mate it up to the filter seat and turn to tighten. No tools are necessary as hand-tight is adequate.

To remove a stuck filter, attach a filter wrench and, using a hammer, gently tap clockwise.

NOTE: The most important lesson is *do not over-tighten* and make sure you use the correct tools.

The following five photographs show various types of filter variations.

Barrus, showing the location of the spin-on oil filter.

Vetus (Mitsubishi), showing the location of the spin-on oil filter.

Beta (Kabota), showing the location of the spin-on oil filter.

Canaline, showing the location of the spin-on oil filter.

Yanmar, showing the location of the spin-on oil filter.

COMMON PROBLEMS AND SOLUTIONS

Following are some of the common problems and their solutions:

- The filter is leaking after replacement. Remove and check that the filter seal is not twisted, broken and that the seat is clear of debris. If possible, use an alternative seal, or, in an emergency, the old filter seal can be turned over and fitted 'upside-down' if in better condition.
- If an aftermarket filter has been used, check that the correct filter has been supplied and that the filter is not 'loose' on the fitting.
- If the oil-pressure warning lamp/alarm activates, this could indicate water in the oil. This will look creamy or 'frothy' and can be a clear indication of a head gasket failure or cracked head, especially if bubbles are also present in the cooling system. Alternatively, check for issues with the oil cooler.

CHAPTER 3

COOLING SYSTEMS

INTRODUCTION

Marine engines are either water cooled, in which case the cooling system serves the same function as a radiator on a standard vehicle engine, or air cooled, as on a motorcycle engine. The main purpose is to conduct heat away and to cool the engine. A water-cooled system transfers heat from the engine by passing water through it. The water can be from an external source, or contained within a sealed system held on the boat. Air cooling is where the action of blowing air through the engine's cooling fins transfers the heat away from the engine. The air-cooling system used to be the most common system found in marine applications, but water-cooled systems are now the preferred method of cooling.

It is vital that any cooling system on an engine is fully operational and maintained regularly. Any blockages of air vents and associated filters can quickly result in overheating engines, just as any failure of the water flow around the system can have serious and sometimes catastrophic consequences. It can result in head gasket failure, the exhaust overheating and even the engine seizing. For this reason, every element within the cooling system must be checked and maintained. It is important to understand fully the system installed on your vessel and the components that need to be maintained.

Tools required:

- screwdriver
- sockets
- impeller puller
- heat gun.

BASIC AIR-COOLED SYSTEM

Air-cooled engines are a slightly different design and incorporate 'cooling fins' to allow the heat to transfer to the air surrounding it, similar to a motorcycle engine. Unlike a motorcycle, which moves the engine through the air for cooling, the boat engine needs the air bringing to it. A significant amount of air needs to pass around the engine to cool it efficiently. Vents are usually positioned in the hull around the engine room, often connecting to the engine casing by flexible ducting. Some vessels will have fans to help circulate the air, or employ 'scoops' to funnel the air into the engine room, or use ducting to transport the air from outside directly to the engine.

In general, an air-cooled engine runs a little hotter than other engines, which can improve efficiency. It also benefits from the fact that with no cooling

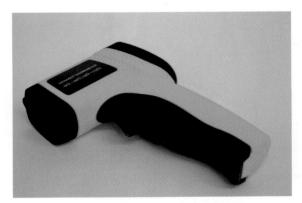

Thermo gun for checking temperature. It can also be very useful when trying to detect airlocks.

pipes full of water, there is no risk of coolant freezing, leaks or scale build-up in the engine. However, an air-cooled engine is generally noisier as there is no dampening of sound. There is also the problem that if the air vents get blocked, or not enough air is circulating, the engine room ambient temperature can quickly increase and result in the engine overheating. The engine cooling fins also lose efficiency if the surfaces are covered by leaves, dirt, debris or are painted; as this is also a fire hazard, regular cleaning is advisable.

The air intakes and outlet vents on the side of the hull present the biggest risk with air-cooled engines, because if something happens that causes the boat to sit lower in the water, these vents will potentially let water into the engine room and can quickly sink a vessel.

Maintenance: Maintaining an air-cooled system is simply a case of keeping all air vents clear, cleaning the engine fins and checking any ducting for cracks or damage. If a fan is used to assist cooling, regularly grease the bearing and check the fan-belt condition and electrical connections.

BASIC WATER-COOLED SYSTEM

Following are explanations of the three main types of water-cooling system.

Raw-Water Cooled – Direct Cooling System

The term 'raw water' refers to the water that surrounds the boat and is used to cool the engine. The raw water can be salt or fresh, and both can be used to cool the engine efficiently. The water is drawn into the engine through a seacock fitting in the hull, then pumped through the engine's water jacket and ports. The heat is transferred from the engine to the water and the heated water exits through the exhaust. The flow of water through the system is driven by a mechanical or electrical water pump and the action of the circulating water is to absorb the heat from the engine to keep it cool.

The raw-water systems are susceptible to scale build-up in the engine cooling system, just like a kettle, which can restrict the flow of water and cause the engine to overheat. This is a common problem for those engines using salt water for cooling. Blockages in the intake are common, so sea strainers or

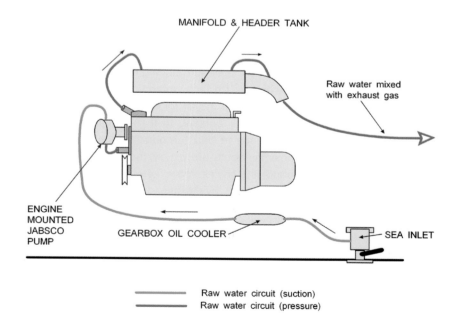

MANIFOLD & HEADER TANK

Raw water mixed with exhaust gas

ENGINE MOUNTED JABSCO PUMP

GEARBOX OIL COOLER

SEA INLET

—— Raw water circuit (suction)
—— Raw water circuit (pressure)

Raw-water-cooled system. Water is drawn up from the outside of the vessel by the engine water pump and is used to cool the engine before being expelled through the exhaust.

mud boxes are usually employed in order to filter the water before it enters the engine cooling system. This system will incorporate sacrificial anodes to inhibit corrosion, which must be replaced at preset intervals (consult your manual or engine manufacturer to ascertain the location). Failure to replace the anodes will result in extreme engine wear and eventual failure.

Heat-Exchanger Cooled – Indirect Cooling (Raw Water)

Most modern marine engines use a separate cooling system to cool the engine. This is very similar to a vehicle system using a radiator to transfer heat from the engine, but in this case a heat exchanger is used. The heat exchanger contains two isolated systems, one for the coolant/water mixture flowing around the engine (a sealed system) and the other in the heat-exchanger jacket, which contains the raw water. These systems still incorporate seacocks and water strainers to filter the raw water. A feature of engines with this system is a second (usually a rubber impeller type) pump. This second pump can be belt-driven or mounted on a Power Take-Off (PTO) flange on the engine.

The heat exchanger (which can be engine-mounted, often as part of the water-cooled exhaust manifold, or remote) has the coolant from the engine flowing through the outer compartment (jacket) and the raw water flowing in copper pipes running through the compartment. Raw water is drawn up

Water pump showing the corrosion and deterioration that can occur on raw-water systems, or be caused by not using inhibitor or antifreeze in an engine cooling system.

Heat-exchanger system – raw water. The raw water is drawn from outside the vessel by an impeller water pump, then travels through the heat exchanger in copper pipes, where it cools the engine coolant that passes around the cooling tubes. The raw water is then expelled through the exhaust.

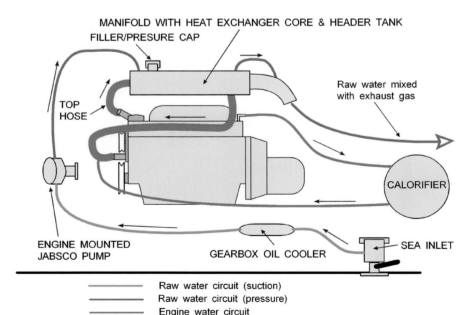

MANIFOLD WITH HEAT EXCHANGER CORE & HEADER TANK
FILLER/PRESURE CAP

TOP HOSE

Raw water mixed with exhaust gas

CALORIFIER

ENGINE MOUNTED JABSCO PUMP

GEARBOX OIL COOLER

SEA INLET

——————— Raw water circuit (suction)
——————— Raw water circuit (pressure)
——————— Engine water circuit

TIP: HEAT-EXCHANGER CLEARING

When water flow is restricted because of scale build-up, the engine begins to overheat. It may be possible to 'acid boil' the scale out of the heat exchanger and continue to use it. The worst-case scenario is that you would have to replace the heat exchanger, but this would be much less expensive than replacing the engine. Salt or limescale can also build-up in the exhaust mixing elbow. This builds up exhaust back pressure, which acts on the raw water and can force the Jabsco vanes down, stopping it pumping sufficiently – as can a delaminating exhaust hose. (Hose collapses internally.)

through the seacock and flows through the heat-exchanger tubes, rather than around the engine as in the direct raw-water-cooled system. The separate engine coolant mixture is circulated around the engine and then travels through the heat exchanger, where the heat is transferred to the cooler raw water. The heated raw water is then pumped out through the exhaust. Please note that there may be anodes

installed in the raw-water part of the system (consult your manual or engine manufacturer for clarification).

The benefits of this system over the raw-water system are mainly that scale build-up is restricted to the heat exchanger, instead of affecting the whole engine. If this does become an issue, replacement of this part is easier and cheaper than a full engine recondition/exchange.

Keel-Tank Cooled – Direct Cooling

A common alternative to using raw water to cool the engine is to provide the water from an enclosed system held on the boat. This is the most common method employed on inland waterways vessels. The hot coolant from the engine is circulated through a keel cooling tank or a series of pipes under the hull, or via tanks built in to the hull where it loses heat to the water and air surrounding the tank before it then returns to the engine.

An empty heat-exchanger shell in the form of a water-cooled exhaust manifold is often used as the coolant header tank. This can cause confusion when trying to identify the type of cooling system employed on the boat

The coolant from the keel tank flows through the engine, through the empty heat exchanger and back to the tank. However, on some engines, like the Beta

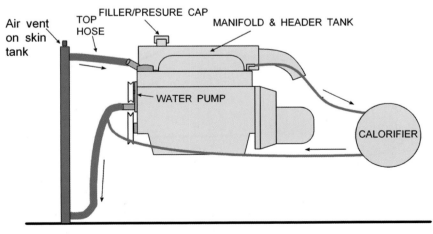

Keel-cooled system. This is a sealed system, which contains the water-coolant mix that travels around the engine before returning to the keel tank.

and Barrus, the heat exchanger can also incorporate the oil cooler for the gearbox and will have hydraulic pipes connected to it. Other heat exchangers can incorporate the engine or the gearbox oil cooler, or charge air coolers (intercoolers) on turbo-charged engines.

Identifying and Removing an Airlock

The most common issue with this system is airlocks, which will cause the engine to overheat. To identify an airlock in a keel tank, feel the tank; it should feel hot at the top and cooler towards the bottom. If there is no variation, there is an airlock stopping the water circulating. Most tank systems have a bleed bolt on the tank to release the air (similar to a radiator bleed, just bigger). If this is not present, run

the engine from cold with the radiator cap removed from the heat exchanger (header tank).

MAIN COMPONENTS ON WATER-COOLED ENGINES

Whether the engine utilizes a raw-water or an enclosed-water system, there are a number of common components that are used in both types of systems and it is important to identify and maintain each element. The relevant sections below will provide guidance on maintenance and are defined for raw-water or keel-cooled systems. We will discuss the following: the seacock; sea strainer (or mud box); hoses, clamps and belts; water pumps; impeller; and thermostat. We will also look at the

1. Locate the coolant tank bleed bolt. This is usually a square bolt or a 17/18mm bolt on the tank.

2. To clear an airlock, undo the bolt using an adjustable spanner.

3. Once loose, it can be loosened by hand; do not remove completely.

4. Allow the air to leak from the bolt hole. Leave it until water comes through, then tighten, much like bleeding a house radiator.

heat exchanger, expansion tank and exhaust-water injection pipe.

Seacock (Raw Water)

The seacock is a through-hull device that allows water to enter the cooling system through the hull. This device has a handle that allows you to shut off the water flow manually or electronically. These handles are used to close off and isolate the system when the vessel is being winterized, or to isolate the supply for maintenance, or if you have a problem such as a loose or cracked hose or clamp.

To maintain the seacock operate regularly and grease to keep from seizing. It is also worth having a wooden plug close by in case of an emergency.

TIP: PROTECTING THE RAW-WATER SYSTEM

It is always worth employing a back-up safety measure. Try to make or obtain a soft, tapered, wooden plug (a bung) the same size as the seacock hole, which can be used to plug the seacock and stop water flow if for any reason you cannot operate the handle. Bags of assorted sizes of bungs can be picked up from most chandlers. Some recommend that a suitable sized bung is located near every seacock for emergency use

Maintenance: Test that the seacock shuts off on a regular basis to make sure it is operable and spray with grease regularly to keep it from seizing up.

Sea Strainer or Mud Box (Raw Water)

The next part of the engine cooling system is the sea strainer. Although called a sea strainer, it is used for any type of raw-water system. The raw water flows through this component, which is designed to filter out debris, sand, leaves and the like before they get to the engine. It can also be found as an attachment to the seacock, but in most applications it is mounted remotely with good access. Prior to checking the sea strainer, it is always advisable to shut off the stopcock, but don't forget to reopen it once you have completed the checks.

Sea strainer, sometimes referred to as a mud box.

Maintenance: There are several types of strainers, but all have a removable filter or screen that should be checked and cleaned or replaced on a regular basis, especially if the engine overheats, you become grounded, or are travelling through water with a lot of weed in it.

- Shut off the seacock.
- Remove the lid to get access to the basket. These usually screw on, but some may have flip-up catches or butterfly screws to hold the lid in place.

- Remove the plastic/gauze basket and inspect, remove any debris and if required rinse out the basket before replacing.
- Check the seal on the lid. This is usually an O ring. Make sure it is seated correctly and in good condition and replace if required.
- Some systems do not 'self-bleed', so once you have emptied the strainer, you will need to refill it with water before replacing the lid.
- Refit the lid and open the seacock. (If the water drains out the seal is faulty.)
- Start the engine and check that water is coming out of the exhaust to ensure the cooling system is operational.

1. Locate the strainer and unscrew the butterfly nut (nut or clamp) to release the lid.

2. Remove the nut completely in order to lift off the lid.

3. Remove the gauze or basket and check for debris. If dirty, rinse out.

4. Remove the gauze or basket and check inside the strainer.

5. Inspect the seal, making sure it is complete and not twisted, before you refit the basket. On some systems you may have to fill the bowl with water before you fit the lid.

Hoses, Clamps and Belts (All Systems)

These are vital to the cooling system and need to be checked periodically. Developing a habit of checking before all long journeys will reduce unexpected breakdowns dramatically. Visually inspect hoses, clamps and belts for wear.

All hoses that are below the waterline should be double-clamped. This will help to prevent water from entering the bilge should one of the clamps fail. Fan belts on some engines drive the water pump and it is imperative that if these fail you do not run the engine, as it will overheat and be at risk of seizing. Checking the condition of belts is covered later in the book.

If you find a corroded clamp/jubilee clip, or a pinched or cracked hose or belt, they should be replaced immediately. Be sure to replace the hoses with the same-size diameter, length and temperature requirements that the manufacturer suggests.

TIP: CLEARING BLOCKAGES

If a hose is blocked and it is a straight section, disconnect both ends and use a pipe cleaner or rod to clear the blockage. For any hose with bends, disconnect both ends and, using a dingy pump or air foghorn, blow through to try to clear the blockage.

Engine Water Pump (All Systems)

On modern engines, this pump is integral. It is belt-driven, with the belt also usually driving an alternator. Whilst they are maintenance-free items, some basic checks can detect early signs of failure. There is usually a 'telltale' hole in the casing behind the pulley that will leak coolant when the internal seal starts to fail. It is also worth checking the pulley for play when checking the belt tension, as this can show early signs of bearing failure, or that the pulley is simply loose on the pump.

The Engine Water Pump

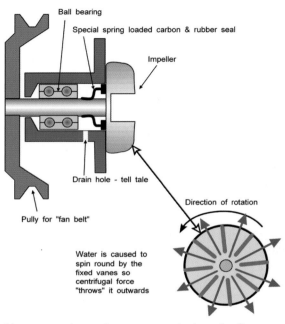

How an engine water pump controls water flow internally.

Check behind the housing to look for signs of corrosion and check for movement on the bearings by testing rotational movement.

ABOVE: Raw-water pump on a BMC engine (PTO type). There are five different types of pump for the BMC engine; note the number of bolt holes and the size of the impeller.

TOP RIGHT: Raw-water pump on a Vetus engine (PTO type).

RIGHT: Engine-water pump on a Beta engine.

Some pumps may have grease or oiling points for the bearings. On older and vintage engines the water pump is often driven by a belt, of the same type as used on raw-water systems.

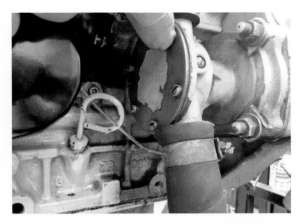

Raw-Water Pump (Raw Water)

The raw-water pump is usually remotely fitted and can be identified by following the pipe-work from the strainer to the engine. These pumps are usually belt- or PTO-driven. Typical types are Jabsco or Johnson pumps, but there are many different manufacturers. When the impeller fails or a blockage occurs in the inlet pipework, the water flow and subsequent cooling of the engine will be affected and can have serious consequences. These issues will all result in the engine rapidly overheating and can lead to exhaust systems overheating, head gasket failure and, in the worst case, the engine dry-seizing. For this reason, the maintenance and familiarity with the location and access of the water pump is essential.

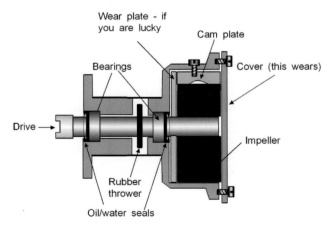

Typical raw-water pump, showing how the impeller pump operates, cut-out view.

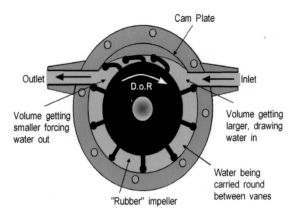

This front view shows how the impeller turns inside the housing and controls the flow of water from the inlet and outlet pipes.

Impeller

Access to the impeller to inspect or replace it will usually require the pump faceplate to be removed. However, if access is an issue the pump may need to be removed to gain access to the faceplate. The raw-water pump is susceptible to a range of problems and should always be checked if the vessel becomes grounded, or is operating in heavily weeded areas. The most common issues are fine sand and silt being pulled in through the raw-water intake (these are abrasive and will quickly cause damage to impellers), or a blockage of the intake will restrict water flow and the impeller will overheat. This can cause the impeller to break up and result in blockages in the pipework.

TIP: PHOTOGRAPH BEFORE DISMANTLING

Use the camera found on most phones to record how things look before you take them apart, or make a quick sketch if you prefer the old-school method. This precautionary measure is recommended for just about anything you need to dismantle.

Maintenance: The impeller is classed as a consumable item and the manufacturers recommend that they are changed every year. The rubber becomes hard and brittle and, depending upon cruising area, can quickly degrade. If an impeller has failed, always make sure you locate all parts, as they will inevitably end up lodged in pipework and cause overheating.

How to locate the raw-water pump:

- Follow the pipe from the sea strainer. Most pumps are fitted directly to the bulkhead, but can be mounted to the engine or part of the engine. Some engines like the Barrus Shire have the pump driven from the front of the crankshaft; typically, these are Johnson or Jabsco pumps.

How to locate the engine water pump:

- Vintage engine cooling systems often use belt-driven pumps, which are straightforward to locate, whilst others will require following the pipework as the pump may be remotely mounted.

How to change the impeller:

- Close the seacock on raw-water systems.
- Remove the faceplate, which is usually mounted with four to six screws. Carefully remove the gasket. The impeller kit usually comes with a new gasket, which should be used when refitting. Alternatively, if the gasket is in poor condition and there is no replacement available, use a non-hardening gasket sealant when replacing the faceplate.
- Check the impeller condition and note which direction the blades are facing.
- Remove the impeller by twisting and pulling with a pair of pliers or a specialist 'puller'. If the impeller has failed and there are parts broken off or missing, you must make sure that you locate all the parts. They may be lodged in the pipework and will cause a blockage, resulting in the engine overheating if not removed.
- Fit the new impeller. The blades on the impeller will always travel in a certain direction; this is

Remove and check the condition of the impeller and which way the blades are facing. If parts are missing, make sure these are located before replacing with a new one.

from the inlet to the outlet via the longest route. It is important that the impeller blades are trailing behind the direction in which it turns. If you experience any problems inserting the impeller, use a cable tie around the blades to compact them, insert the impeller part way, then cut the cable tie and remove before fully inserting and fitting the faceplate.

- Replace the gasket and fit the faceplate, making sure not to overtighten the securing screws. (Ensure the information plate is facing outwards; if it faces inwards, the information on the plate will wear away.)

- Open the seacock on raw-water systems.
- Start the engine and check that water is coming out of the exhaust. As a secondary check, look for leaks from the water pump gasket and feel the front plate; any heat could indicate that the pump is running dry, so remove the plate and check operation.

2. Water may start to escape before all of the screws are removed, so it is useful to leave one in loosely in order to swing the plate away from the face.

1. Remove the screws around the faceplate.

3. Remove or swing the faceplate clear to expose the impeller.

4. Using a screwdriver, gently ease the impeller out of the body, noting which way the fins are bending.

6. To fit the new impeller or refit the old one, mate up to the housing and twist in the direction the fins need to travel, then push gently.

8. Replace the faceplate, making sure that it is facing the correct way, as any information on the face will be erased if facing towards the impeller.

5. Remove and inspect the impeller. If any parts are missing, these must be located and then the impeller replaced.

7. Before pushing the impeller home, check that the fins are rotating in the correct direction.

Thermostat (All Systems)

The thermostat restricts the flow of water to ensure that the engine is running at its optimum temperature. The thermostat's operating temperature is matched to the engine and is often lower in marine applications than for typical automotive or plant applications. The thermostat will be marked with its operating temperature, which dictates when it starts to operate, and a different temperature value that it has to reach for it to be fully open. Independent of this, the thermostat has a small bleed hole that allows water to flow even when the valve is closed. Many incorporate a second valve, which controls flow through the engine's internal bypass port. This valve resembles a 'penny washer' attached to the base of the thermostat.

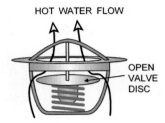

HOT WATER FLOW

OPEN VALVE DISC

This end must be inside or towards engine

Thermostat operation, showing how it controls water temperature (single valve).

When an engine is cold, the thermostat is fully closed and the engine coolant flows around the engine, but not through the cooling system. This is designed to allow the engine to get to running temperature. As the engine starts to reach running temperature, the thermostat begins to open and allows some of the engine coolant to flow into the cooling system, enabling the temperature to be 'controlled'.

Once the engine is at its rated running temperature, the thermostat adjusts its position to reduce or increase coolant flow, dependent upon engine running. It then continuously controls the engine's operational temperature to ensure that the engine is running at its optimal temperature at all times. It is therefore important to check the thermostat on a regular basis to ensure it is operating correctly. Overheating can occur when the thermostat is open fully, but the amount of heat being created is greater than the cooling system can deal with. Or it can happen when the thermostat fails to open and the engine temperature continues to increase. There are many other possible causes of overheating; all will require investigation and should not be ignored.

Overheating can damage thermostats; if they fail due to overheating they usually fail in the open position (making the engine run cooler). They can occasionally fail closed, causing the overheating to recur. They should always be tested and replaced if necessary.

One of the more interesting causes of overheating can be running the engine without a thermostat fitted. This is only an issue with double-valve thermostats, where the internal bypass port is constantly 'open' once the thermostat is removed. This port allows up to 50 per cent of the coolant to recirculate, causing a 50 per cent deficiency of the cooling system. This usually causes a problem under load, as when idling little heat is produced. Always seek guidance or check your owner's manual prior to attempting to run with no thermostat fitted.

A new thermostat and a failed thermostat. Note that the failed one is in a permanently open position. (The new one is a dual valve.)

TIP: HOW TO TEST A THERMOSTAT

Quick and easy testing of the thermostat can be achieved by placing it in a pan of water, which is then brought to the boil. You will need a thermometer (or heat gun) to monitor the temperature and note at what temperature the thermostat begins to open. Remove it from the water and ensure that it closes again. This will confirm that it is operational. Details of the thermostat's operating characteristics can usually be found on the thermostat, workshop manual or by searching the part number on the Internet.

Maintenance: To locate the thermostat housing, look directly above the engine water pump. There should be a casting with an elbow protruding from the top of the engine. This is the thermostat housing position on the majority of vessels. However, it may also be located in the side of the header tank; if this is the case, the main coolant hose will attach here instead of at the bottom of the header tank. The housing will be secured using two to three nuts/bolts, which will need to be removed to gain access to the thermostat beneath the housing.

1. Thermostat housing on a BMC.

2. Thermostat housing removed to access the thermostat on a BMC.

3. Thermostat housing on a Beta.

Following is the procedure for removing or changing the thermostat:

- On direct raw-water-cooled engines, close the seacock.
- Loosen the pipe clamps to the thermostat housing; some coolant may leak out, so have a cloth handy to catch it.
- Remove the housing bolts and lift the housing clear of the engine (or as far as any connecting pipes will allow). As the housing and connecting pipes will be full of coolant, it is always best to have a container or cloth ready to catch the coolant.
- An alternative on true raw-water systems is to drain the water pump prior to undertaking the above procedure, but in my experience access to the pump is usually difficult and removal and replacement of the faceplate can result in rounded screws or damaged gaskets. (On heat-exchanger or keel-cooled systems, draining the water will not be possible.)
- Inspect the gasket for any signs of wear and replace if required.
- Lift out the thermostat and test operation.
- Replace the thermostat if required or refit the old one, then refit the housing, remembering to tighten up the pipe clamps before finalizing. Please note that if in doubt, it is always best to replace the thermostat. It may still be operating, but not efficiently, and this distinction is difficult to observe.

- Top up the coolant (using any saved from earlier if applicable).
- On direct raw-water-cooled engines, open the seacock.
- Start the engine and run up to temperature. Feel the housing – it will be cool, then will become hot once the thermostat opens.

3. Using a spanner, loosen one of the bolts, then use fingers to unscrew.

4. Remove the bolt completely from the housing and place it in a container.

1. Locate the thermostat housing (Vetus).

2. Undo the other bolt, remove it completely from the housing and place it in a container.

5. Locate the jubilee clip holding the main hose to the thermostat housing and loosen. In this case, this is enough so that the thermostat housing top can be moved. In other cases you may have to remove the pipe connected to the housing.

6. *This thermostat housing can be lifted (pivoting on the part connected to the rubber pipe), so therefore you can prise it open by applying pressure in the right area. On others, the whole lid must be removed.*

7. *The thermostat housing lifts, giving access to the thermostat, which is seated in a depression. Once open, remove the thermostat.*

8. *Inspect and test the thermostat, then refit and replace all bolts and tighten the jubilee clip.*

Heat Exchanger (Indirect Cooling)

The heat exchanger is used to transfer heat from the engine coolant to the water flowing through it. It does this by passing the raw water through a chamber full of small copper tubes called a tube stack, which is contained inside the heat-exchanger body. The hot engine coolant flows around the tubes inside the body. The raw water in the tube stack is cooler than the engine coolant and therefore absorbs the heat through the tubes. The heated raw water is then transported to the exhaust. The heat-exchanger capacity and efficiency should be matched to your engine's specifications. If the heat exchanger is suspected to be blocked or to have failed, it can be removed for repair or replacement. This is a relatively easy process, but is rarely required on the inland waterways.

The tube stack slides inside the jacket and is sealed by O rings at each end. They generally sit between the end covers without further restraint, but are sometimes clamped at one end only. The fitting is the same for standalone and for those incorporated into the water-cooled exhaust manifold. Many engines fitted with combined water-cooled manifold and heat exchanger have rubber end caps over the ends of the tube stacks. These are approximately 75mm in diameter and will both have pipework connected to them. These rubber caps are usually referred to as Bowman or Polar end caps, after their respective manufacturers.

Maintenance: The rubber end caps can become damaged, split, swollen or brittle, depending upon

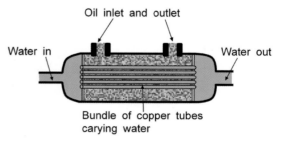

Heat-exchanger internals. Copper pipes carry the cooling water.

1. Heat exchanger with Bowman end caps on each end.

2. Check the rubber by squeezing. It should be firm, but check for any distortion, cracks or rubber degradation.

3. Oil cooler with Bowman end caps. These should be checked by squeezing and should be firm.

Examples of Bowman and Polar end caps.

age, overheating and usage. These should be regularly inspected and it is always advisable to carry spares. Each end cap has a unique number embossed on it and the parts are available online or from most chandlers.

On some keel-cooled engines it is common to find that the heat exchanger is combined with an expansion (header) tank to allow for the expansion of liquid that can occur on sealed systems. The heat exchanger's shell is used for the water-cooled exhaust manifold and 75mm caps are used to block the holes. In most cases, they do not have pipe-work attached to them as the tube stack has been removed.

Heat-exchanger shell on a Beta engine with caps on the end. This may also contain the oil cooler for the gearbox.

Expansion (Header) Tanks (Heat-Exchanger and Keel-Cooled Systems)

The expansion tank is an essential part of the sealed cooling system, as it allows for the expansion of the cooling fluid (as this heats, it expands and its volume increases). This small tank is often part of the water-cooled exhaust manifold (and heat exchanger if fitted), or can be a separate tank that is mounted remotely; both perform the same function to allow for any fluid expansion. On all expansion tanks there is a cap, which is equivalent to a radiator cap, and this will have a pressure rating that should match your engine pressure specifications. The rubber seal should be checked regularly and the radiator cap should be changed periodically.

NOTE: An overflow tank is often provided where the expansion of the coolant cannot be contained in the header tank, typically when the system has a very large volume of coolant. It is an unpressurized vessel connected by a hose to a connection on the side of the filler neck below the pressure cap. The hose must either reach into the bottom of the tank, or be connected to the bottom to allow the coolant to return to the engine as it cools.

Maintenance: Locate the expansion bottle, remove the cap and check the seal.

An external expansion tank should always be mounted higher than the engine, to allow for air to escape and not become trapped in the system. This is also true of the connecting pipework, which should also have no tight bends or dips.

Exhaust-Water Injection Pipe (Raw-Water and Heat-Exchanger Systems)

The majority of raw-water-cooled systems will incorporate a water-cooled exhaust system. Having passed through the engine or heat exchanger, the raw water is transported away from the engine and is injected into the exhaust via the injection bend. The injection bend is subject to a mixture of gases and water, which can cause internal corrosion; this component should therefore be checked for scale build-up, damage or cracks, and cleaned or changed if any are present. It is worth noting that most failures of the injection bends are on welds, but they can occur anywhere and there is no evidence that different metals or designs provide any benefits.

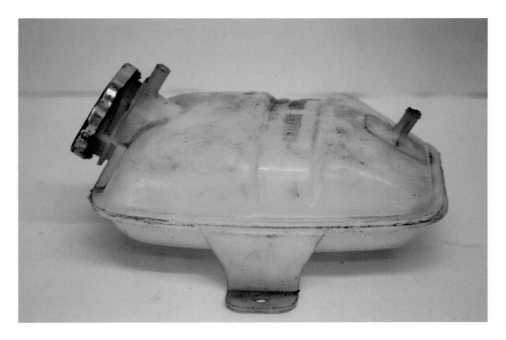

Expansion bottle.

The water and exhaust then passes through rubber hoses to the water baffle (usually plastic), which acts as a silencer before exiting the boat via a skin fitting. Failure of the raw-water pump can cause plastic parts of the exhaust to melt.

BLEEDING THE COOLING SYSTEM

Whenever the cooling system is drained, has experienced a leak, or a component is removed, it will be necessary to bleed the system and remove any air in the pipework. Depending on the arrangement in your boat, the type of fit out and complexity of interconnecting systems will dictate how easily this is achieved. The approach and methodology is the same, but with a more complicated system it will be necessary to look for 'trouble' spots where air can get trapped and may require a bit of manual assistance.

Air can become trapped in heat exchanger or keel/tank cooling systems, especially if a calorifier is involved. If there are any hoses above the header tank, this will virtually guarantee an airlock. Another risk area is where cooling hoses are installed to form an inverted U, as the air will sit at the highest location and although some water may get through it will restrict the flow. If you have a section that is difficult

to bleed and cannot be rerouted, one option might be to invest in a cooling hose with a bleed point. The best practice to employ is for all hoses to slope downwards and be below the level of the header tank.

Bleeding the system:

- Once all the components have been refitted, top up the system slowly to minimize the amount of air trapped in the system.
- Remove the radiator cap and run the engine. Bubbles should appear; rev the engine until these clear, or until the thermostat opens – the water will steam or circulate in the heat exchanger. Be careful to keep clear of the tank, as a blown head gasket or large airlock can cause super-hot water and steam to blast out of the cap unexpectedly.
- Air trapped in the calorifier can be extracted by removing the calorifier return hose from the engine. Block the engine connection with a thumb whilst running the engine on tickover. After a while, water should gush from the hose; replace the hose and top up the cooling system. *Do not attempt this when the engine is hot.*
- Rev the engine for a minute or two and replace the cap.

CHAPTER 4

ENGINE ELECTRICAL SYSTEMS

INTRODUCTION

The aim of this chapter is to outline practical guidance and knowledge to enable you to test and diagnose the most common electrical systems in your vessel and also to provide some background to electrical theory.

It is important to be able to identify the components that make up each system. To facilitate this, a glossary of common components, along with an explanation of what they do, has been provided. Not all components will be included and you may find some that have not been listed. However, the aim is not to make you an electrical expert, but simply to provide the basic tools to allow you to trace faults and implement repairs.

Understanding the tools needed to deal with the variety of components and cables that make up your systems is essential if you are to be able to carry out maintenance and implement repairs. This chapter includes an overview of some of the basic kit required.

It is not essential that you read the theory section that follows the practical section, as the practical section can be followed without the need to appreciate fully the complexity of electrical theory. I have always found that some people 'get it' and others simply do not, but one thing that is guaranteed is that understanding the theory will help if you are upgrading or changing your systems.

With any installation or maintenance work involving electrical systems, it is important to apply basic common sense and, if undertaking major changes, to work within the guidelines set out in EN ISOs 10133 [low voltage DC installation] and 13297 [low voltage AC installations], both of which can be found in the BMEEA Code of Practice (available from the British Marine Federation).

Undertaking electrical maintenance, fault-finding or installation can be tricky and should not be undertaken without some guidance, practice and, where possible, training. Most electrical installations, unless new, can be very difficult to get to grips with, due to the number of modifications, additions and rewiring that may have been implemented. In some instances, it is recommended that a professional is employed to tidy and test your system first and to take you through the basic operation before you attempt to undertake any work yourself.

TOOLS AND TEST EQUIPMENT

Using the correct tools and understanding when, how and why to use them are important factors when working on electrical systems. Using the correct tool is guaranteed to save time and give you confidence in your diagnostic capabilities as you develop knowledge and skills. It is possible to spend a small fortune on hand tools, but all that is required is listed below.

Tools required:

- Set of electrical screwdrivers
- Medium sized pair of side cutters
- Pair of wire strippers
- Ratcheting type of cable crimper
- 8, 10, 13, 17mm spanners
- Multimeter.

Optional:

* Ferrule crimper and ferrule ends.

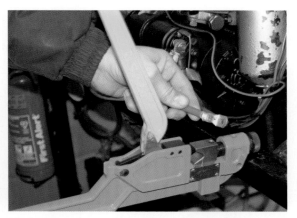

Hex crimpers are the recommended tool for crimping heavy duty cable.

Ferrules are used to provide extra-secure connections. This is especially important where multicore cable is used and in environments where vibration is present. The crimping tool is used to put a ferrule on the end of a cable or wire. There are many different sizes to accommodate wires of different gauges.

Test equipment required:

* An RMS multimeter (it must be an RMS meter so that AC voltages can be measured correctly, particularly those generated by invertors).
* A clamp-type multimeter. This may remove the need for an RMS multimeter, as it will perform both types of measurement, current and voltage. However, these are more common for AC work and some do not have the function to measure DC, so make sure you check before you buy (most engines' electrical circuits are DC).
* A hydrometer for testing battery fluid.

One of the most important and beneficial skills is being able to attach new ends (lugs) to large cables, as this is one of the most common causes of electrical problems. However, different cable sizes require different tools. Before discussing what tool is required for work with cables of a larger cross-sectional area, it is important to understand the different methods that you may encounter.

It is now an industrial standard to crimp all cable lugs and has been so for some years, but it is still common to come across installations where the cable lugs have been soldered on to the cables. There are a number of different styles for crimping a cable lug on to a cable, but the accepted standard is to use a hex-style crimp.

There are many other styles of crimps, but these can cause poor conductivity between the lug and the cable. Also, some of the other types of crimping tool will cause damage to the conductor held within the lug.

If you want to be prepared to work with heavy duty cable, typically ranging from 10mm^2 to 120mm^2, then additional crimping tools will be required, as a small hand crimper is only designed to be used on cables up to 6mm^2. The heavy duty cable tools will be required for battery, starter and some domestic circuits.

If you are considering buying a crimping tool for the cable sizes mentioned there are a number of suppliers that can be found on the Internet selling hydraulic cable-crimping tools at very reasonable prices. There are also hand-operated hex crimping tools available.

Standard multimeter.

Clamp-style multimeter, more commonly used on AC circuits. It is important to check that the one purchased is for DC systems.

MULTIMETERS

Electrical installation and maintenance will at some point require the use of some form of test equipment. It is important to understand which type of multimeter should be used and how to use it correctly. Multimeters are measuring instruments that perform a range of functions. They make it possible to measure the properties of a circuit for diagnostic purposes.

- Always refer to operating instructions before using a multimeter and ensure that it is capable of reading what you are measuring.
- You must select the relevant range that is slightly higher than the values you know or expect to be measuring.
- If you are not sure what range to use, always start with the highest one and work down.
- Whenever the instrument is not being used, it must be switched off in order to preserve the internal battery. Most modern multimeters will turn off automatically after around 10min.

Multimeter functions:

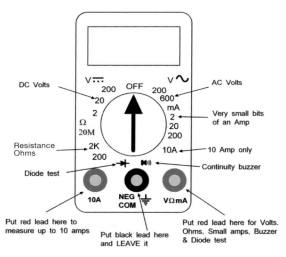

Typical settings and functions of a multimeter.

- Continuity settings. This setting will normally provide an audible noise when the probes are touched together, confirming that there is a continuous circuit. If the audible alarm does not sound, it indicates that the wire is broken or you are not connected to the same wire at one end.
- Voltage settings. Use the DC setting and always choose a measurement range carefully and to match the reading you are expecting (for example, VDC-20 for a 12V system).
- Resistance and diode function. This can be used instead of a continuity setting; it will not be audible, but will provide a more accurate measurement.

- Current – standard current setting. This is useful to see what the consumption and rating of a component is.

COMMON COMPONENTS

Following are the typical components found in the electrical systems.

Engine Panel

The instruments on your dashboard will dictate the level of information and detail to which you have access. Typically, this will include the following items.

The engine panel may contain voltage, amp and rpm, oil-pressure gauges, ignition and warning lamps.

- Ignition/charge light – when illuminated it indicates that the engine is not charging and could be a fan belt or alternator failure. It should always

be investigated if illuminated (it should always be illuminated until the engine is started, although a rev may be required initially).

- Oil light – fed from the engine oil-pressure sensor and illuminated when oil pressure is low. Sensor failures are a common cause of breakdowns and the best way to check is to wait for the engine to cool and then restart it. If the indication is immediate, it is a sensor failure not an oil-pressure issue. Alternatively, if you know someone with the same engine you could borrow a temperature sensor to test if the symptoms persist. Depending on how well you know your engine, it may be possible to judge from the engine note. If the engine sounds like it always does, chances are the sender is at fault. However, it is a good idea to avoid running the engine until the sender is replaced to be absolutely sure.
- Voltmeter – indicates voltage level of batteries (can be connected to domestic bank or starter battery, but always best to identify).
- Rev counter (tachometer) – indicates the engine rpm.
- Amp meter – indicates the amp charging rate (you may have different gauges for domestic and starter batteries or alternators). If there are two alternators and only one gauge, it is always worth checking to which one it is connected.

Ignition Switch

This key-operated device can come with any number of positions depending on the system

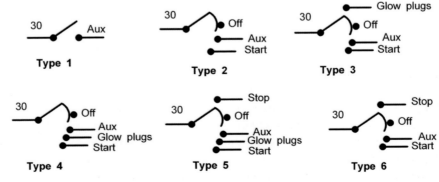

Positions on the ignition barrel, showing the number of combinations that an ignition barrel can have.

30 = Main battery feed into switch

employed. However, most will have a position that includes: Ignition On; Start (engages starter motor); Stop; and Heat (heats the heater/glow plugs). The heat position is usually a sprung return, so that you have to hold it in position, then it springs back to the normal position.

Warning Buzzers

These are typically linked to some of the items on the panel so that there is an audible alarm for warnings for bilge pump operation, low charging and low oil pressure. There may also be visual indications. When the ignition is initially switched on or off these alarms will sound, because when the engine is at rest there is no oil pressure, alternator charging and so on. This is perfectly normal, although if they continue to sound once the engine has been started this should be investigated.

Starter Battery

This is common to all systems – the battery supplies the power that allows the engine to start. Batteries come in many forms and ensuring that you have purchased a 'cranking' battery for engine starting is extremely important. The starting battery provides cranking power, which means that it delivers a high power output for a very short amount of time and is designed to do this repeatedly without deterioration.

Many suppliers sell dual-usage batteries, which can be used for both domestic and starting applications. However, these will not have the life of a true starter battery. When purchasing an engine battery from motor factors, if you request a starter or cranking battery for a Ford Transit van, this will usually ensure that you obtain a suitable engine battery. Further information on types of batteries can be found later in this chapter.

Battery Master Switch (Isolator)

This switch is a common component to all systems and is installed to allow you to isolate the electrical systems; this stops batteries from draining/discharging should there be a drain in the system or some electrical device, such as a light left on.

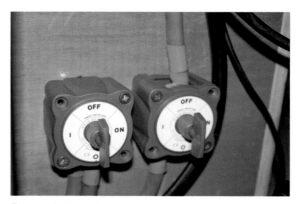

Battery isolators.

TIP: NO POWER FROM THE BATTERY

The battery master switch does suffer from 'furring' caused by moisture build-up on the contacts, so it is always worth spraying with an electrical spray when leaving for periods of time, as well as moving the switch from one position to another when you return, in order to help clean the contacts.

(Normally the bilge pump is connected directly to the battery, so that it can continue to operate even when the battery is isolated.)

Starter Motor

The starter motor is an electric motor that spins the engine in order to start it. The key to the engine starting is the speed at which it spins and this is dependent upon adequate power being supplied to it. The power in the batteries, a good electrical connection via the wires and a good, sound earth connection are all essential, along with the starter's ability to engage the flywheel teeth inside the bellhousing.

The starter contains a starter solenoid. In older models, this can be a separate solenoid and if this fails it can be replaced independently of the starter.

Starter motor components.

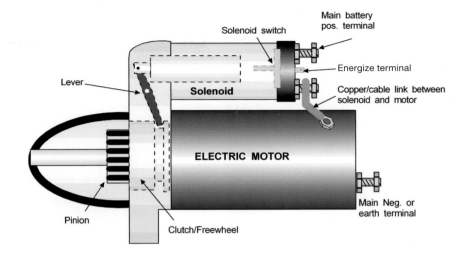

To remove the starter motor, take a photo of the connections on the reverse, isolate the batteries, then disconnect the electrical connections and unbolt the body.

Starter Relay

This component is sometimes used when the starter is located a long distance from the starter button or ignition switch. It is an electronic relay that is used to boost the current to overcome poor starting, which can occur in these situations. If not installed from new, these can be a source of loose connections as deterioration of wiring can occur over time.

TIP: STARTER MOTOR CHECK AND EMERGENCY START

If your engine is not starting and you believe the starter solenoid is at fault, or to start the engine without the ignition, connect the positive and negative terminals (on the solenoid) by using a jump lead or a heavy duty screwdriver. This will 'short' the terminals on the starter motor, bypassing the solenoid and energizing the starter motor. *Be aware that the full battery amps are flowing, so be extremely careful and remove the connection as soon as the engine starts.*

Alternator

The alternator is used to generate electricity to recharge the batteries and to supply power to other ancillary components. The alternator is driven by the fan belt, which is attached to the engine crank shaft. Keeping the fan belt in good order will ensure that the alternator continues to deliver power to the batteries.

The alternator is operating in most cases in a damp environment and therefore is subject to failure more often than one on a vehicle. The alternator fitted to a marine engine may have been retrofitted with a battery-management wire that provides information to a battery-management system. However, it is worth noting that this wire will usually invalidate the alternator warranty and it is not a requirement for the system to charge efficiently, although it does provide useful information that helps to maintain the batteries. Further information on sizing of alternators and calculating charging rates is provided later in this chapter.

Cold Start Relay

Most diesel engines employ heater plugs and these have the capacity to be heated electrically to assist starting in cold weather. The cold-start relay is usually incorporated within the ignition system, where you have to hold the ignition in the Heat position. This will activate the relay; alternatively, you may need to press a button for a few seconds to achieve the same outcome.

Stop Solenoid

On the majority of older inland vessels the engine is stopped by operating a stop lever, which is connected to a small cable that cuts the fuel supply to the injection pump. However, on some engines an electromagnetic stop solenoid is used instead and this is operated from a separate stop button, or occasionally from the ignition switch. This sends an electrical signal to the injection pump to stop and requires the button to be held in until the engine stops. However, if the cable seizes or fails, or the relay or electrical signal fails, you will not be able to switch off the engine. It is important to know which system is in operation on your vessel and to carry a spare relay or a spare cable, especially if the vessel has had a long life.

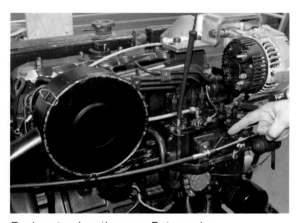

Engine stop location on a Beta engine.

Engine stop location on a Vetus engine.

Engine Manual Stop

The engine manual stop cuts the fuel to the engine and is incorporated within most engine designs, so that the engine can be stopped if other systems fail. It is always worth finding out the location of the manual stop, as not being able to turn off the engine is a very common cause of breakdowns. While this can be very distressing, it is in fact easily resolved.

Close-up view of the stop lever on a Beta engine.

Engine stop location on a newer Vetus engine.

Engine stop location on a Canaline engine.

TIP: COMMON ELECTRICAL FAULT

Most engine wiring is contained within a loom and in order to help installation and testing it is usual for a block connector to connect sections. These can suffer from damp ingress, furring or occasionally the meltdown of an internal contact. This will typically result in the engine not starting, or not turning off. If this occurs, locate the main loom and block connector, pull back the rubber cover and unplug, inspect and spray with moisture-repellent spray and push together, then try the engine to see if the problem is resolved.

Engine Fuse

All vessels should have a fuse located on the positive wire to the ignition switch, which is there to protect the engine circuits from overload or short circuits. Some vessels do not have one fitted, but it is always recommended. If you find you do not have one make it a priority to install one. If the fuse fails the engine will appear to be 'dead'; although a quick check will confirm the cause, remember that simply replacing it will not address the cause of the fuse blowing. It is always recommended that a circuit check is conducted to look for wiring issues or connectors coming loose and so on; if nothing is found and the fault occurs again it is worth calling a specialist in to trace the fault before it develops into a major issue.

Occasionally you may also find your vessel is fitted with a fuse for the batteries too, this is to protect the engine from short circuits causing a wiring loom melt down, however the battery isolator provides a level of protection from this type of event (if it is used).

With some background on test equipment and an overview of the components used in the circuits, you are now in a better position to move forward and look at some of the systems used on your engine's electrics. In the next section we will look at the common circuits and how to test them.

Electrical loom block connector disconnected for inspection.

Electrical loom block connector.

PRACTICAL FAULT FINDING AND CIRCUIT KNOWLEDGE

There are a whole host of electrical circuits employed on boats and it would never be possible for any book to cover the complex nature and different types of installation that can be found on all vessels. However, the aim in the following section is to enable you to gain an understanding of each system, the typical components that are found in them and the methodology that should be applied to test them. The electrical theory follows this section and can be referred to whilst following the electrical testing procedures.

Armed with this knowledge and guidance, testing and diagnosing issues within your systems should be possible. First, we will concentrate on the engine electrical systems that enable the batteries to be charged and the engine to be started. We include ignition systems, electronic stop and the components that are common in many of these systems.

As with other chapters, it is important to identify which components are installed in your vessel and to read up on how these interact with each other. Where possible, this chapter will guide you through the best practice and the key elements that should be installed. If you identify some major items that are missing from the system it is worth considering bringing in a specialist to rectify the issue and to help familiarize you with your system. This will ensure that you can both maintain and troubleshoot competently.

On the inland waterways, it is very rare to find a system installed that comes with the full wiring schematic, even though it is a requirement of the Recreational Craft Directive (RCD) that came into force mid-1998 that all owners' manuals must incorporate a complete set of wiring diagrams. New engines will be supplied with wiring diagrams detailing standard arrangements, which have to be integrated into the boat's pre-existing systems. Unfortunately, most engine and boat builders are unable then to modify the existing schematics. However, tracing your systems and checking them will allow you to familiarize yourself with your wiring and circuits, and if you have the time it is definitely worth drawing the system for ease of reference and also in case you decide to add to or modify the existing systems.

We will start by splitting up each electrical system into separate smaller systems. This should make it easier to understand and identify components, as many people get overwhelmed by the mass of wiring they see. However, the reality is that they are all separate small circuits that rarely interact with each other, making it much easier to ascertain where a fault is present. Please note that this book refers to power as a combination of voltage and current as it travels through the system – this is not consistant with electrical theory, but is used to assist in understanding your systems and circuits.

For the purpose of locating and understanding each electrical system, the following section has been split into four 'primary' systems, which have then been simplified and split into two subsystems.

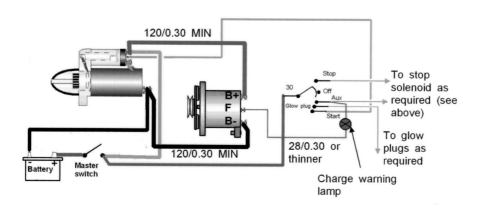

Engine electrical system, showing the main components and minimum gauge wire to use.

Note that your own system may include additional circuits, but from these descriptions you should be able to identify each system clearly and then be able to use the techniques learnt to understand how the circuit works and what its function is. We will be looking at the following electrical systems: engine starting system; charging system; heater plug system; and stop solenoid system.

NOTE: Every electrical system is different and the tests detailed in the following sections are only suitable for positively isolated circuits.

ENGINE STARTING SYSTEM

The starting system has been split in to two subsections, which we will look at separately and will be referred to as the 'main' and 'ignition' systems.

Engine Starting System (Main)

The main subsystem refers to the larger cabling that carries the 'heavy' currents required by the starter motor to turn the engine over. If you encounter faults such as a 'clunking' or 'clicking' starter motor, or experience a 'lazy starter', or a lack of power to the ignition panel, this system is most likely at fault. It is worth noting that dirty, loose or poor connections can cause all of the above symptoms, or result in a complete power loss when a system is put under load, for example when trying to start the engine.

Description of diagram:

- Power is generated at the battery and supplies the circuit.
- Power travels through the battery cable to the isolator.
- Turning the ignition engages the solenoid directly, or activates it via the master switch.
- Power travels through the cables to the starter solenoid.
- Passing through the solenoid into the starter motor, the power finally travels back to the battery through the main earth cables.

Circuit testing: To test and trace a fault on the starting system using an electrical tester, undertake the following tests to identify the location of a fault. Note that the engine does not need to be running to perform this test.

Set the multimeter to the Voltage DC setting (V--) choosing a voltage range just higher than you are measuring. Take a reading from the battery to provide a base voltage level for reference. Please note that a battery which has less than 11.8V present needs to be charged before any further measurements are taken.

- Take the negative (black) probe, connect it to the negative terminal of the battery (usually marked –) and leave it in position during the whole test.

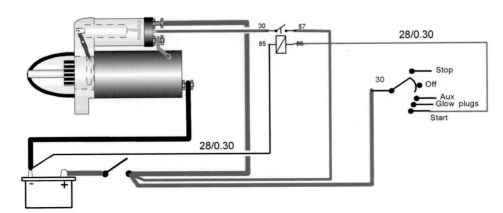

Typical engine starting system (main). Once the battery isolator is closed, the ignition controls the supply of power to the starter solenoid.

Where possible, use crocodile clips, as these will provide a secure connection (and a free hand) whilst you test the circuit.

- Take the positive (red) and place the probe on the back of each of the isolator switch terminals, taking a reading from both. There should be a voltage present at each terminal when in the on position and this should reflect the voltage of the battery (unless the isolator is in the off position). The readings should be within 0.2V of one another. If there is a larger difference, this would indicate a faulty isolator or connection, depending on where the difference has been measured. (The difference in the measurement is referred to as the 'volt drop', that is, more resistance is being experience, therefore the voltage reading is lower.)

- Take the positive probe and place it on the main terminal on the back of the starter motor/starter motor solenoid. This can be identified by the large positive (red) battery cable, which will be connected to the terminal on the back (or the terminal that has no connecting wire/bar to the motor body).

To test the isolator, place the positive probe on the terminals and take a reading. There should be no more than 0.2V difference between the readings from each terminal.

TIP: NO POWER

Isolator contacts can get 'furred up' or damp if left for periods of time, if your isolator is not operational, switching through its positions will usually clean up the contacts and restore it to working order (in the short term).

To check the voltage to the starter motor, place the positive probe on the terminal that the main positive cable is connected to (usually red or brown), with the negative connected to an engine earth.

- If there is no voltage present or there is a difference in the voltage measurement, this would indicate a faulty connection between the isolator and the starter motor, or a fault in the cable. Try tightening and cleaning all terminal connections, then retest. If there is voltage present (which there should be in a healthy circuit and with the battery isolator on), you will now need to perform a check on the negative side (the return – sometimes referred to as the earth) of the circuit.

- Set the multimeter to its continuity setting, then place one probe (either one) on the battery negative and one on the starter motor negative terminal, metal plate or the body of the motor (ensure that it is touching bare metal). The audible signal should sound, the reading will indicate 0 and there may be an audible signal; if not, this indicates a bad earth connection in the circuit. Trace and check all connections on the negative cables especially the main engine earth, tighten and clean any loose connections, then retest.

Using the multimeter continuity function, place the probe on the negative terminal on the back of the starter motor, or on the metal plate. If there is no audible sound this indicates a bad earth connection.

Engine Starting System (Ignition)

The ignition subsystem refers to the smaller cabling that powers the ignition panel and activates the starter motor solenoid. If you encounter faults such as dim or flickering ignition lights, or a starter motor that begins to turn then cuts out intermittently, this system is most likely at fault.

It is worth noting that any loose or corroded terminals will give the above symptoms. You will generally find on older boats that this is the system that has had the most modifications and that the colours of the wires do not always match, or are of incorrect cross-sectional area. To ensure you identify all components associated with the circuit, it is best to trace these systems manually. This can be accomplished by feeling the wire from its start point to its finishing point, or by using the continuity test function on a multimeter. Remember that the most common colours for power cables are red, brown or white with a black stripe. Where incorrectly sized wires have been used, replace with the correct sized wires using the calculations in the next section. You can also disconnect a suspect cable and use a temporary one to check if it is at fault (jump cables are a great tool for this purpose).

Description of diagram:

- This circuit can start at any main positive point, an isolator switch, battery and so on. In this diagram, we begin at the starter motor solenoid.
- Power travels to the ignition switch via the cables.
- Power passes through the ignition switch.

Engine starting system (ignition). On this starting system, the ignition is connected directly to the starter solenoid when the ignition is in the start position.

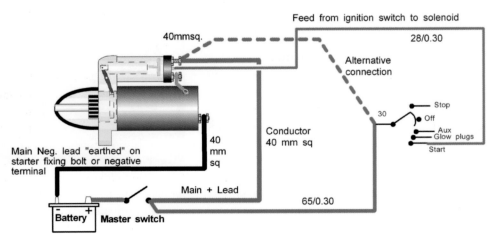

- The circuit is completed when the power returns back via the ignition switch to the starter motor solenoid, via the cables.
- The circuit is completed with the power travelling through the engine to the main negative cable, then returns back to the battery.

Circuit testing: To test and trace a fault on the starting systems using an electrical tester undertake the following tests to identify the location of a fault. The engine does not need to be running to perform these tests, but the ignition must be on.

Set the multimeter to the Voltage DC setting (V--), choosing a voltage range just higher than you are measuring. Take a reading from the battery to provide a base voltage level for reference. Please note that a battery that has less than 11.8V present needs to be charged before any further measurements are taken.

As there are so many variations and modifications on ignition panels, it will not be possible to provide instruction on all checks. Therefore this section concentrates on identifying if there is an issue present on the ignition system. It is recommended that if a problem is found you seek assistance from a professional unless you have more in-depth experience or training. This test requires you to identify a number of components and the terminals required in order to pick up the correct readings.

- Take the negative (black) probe, connect it to the negative terminal of the battery and leave it in position during the whole test. Where possible, use clips as these will provide a secure connection whilst you test the circuit. If you cannot reach the battery use any bare metal on the engine/bell housing.
- Identify the excite wire for the starter solenoid. This is the wire connected to a small spade terminal close to the main positive terminal on the starter. The excite wire provides an 'initiation' voltage, which, in simple terms, energizes the solenoid. This in turn pulls the starter motor gear into engagement with the flywheel teeth and

supplies power to the motor. Take a note of the colour of this wire.
- Disconnect the wire, by pulling gently on the connector; it should come away relatively easily. If you are having difficulty, spray with a penetrating oil and allow time for it to loosen the connection. Once removed, place the positive probe into the connector attached to the wire. Place the multimeter in a position where its display can be read, then turn the ignition key to the start position.

Locate the starter solenoid terminal, then identify the excite wire for the starter solenoid and disconnect it at the terminal.

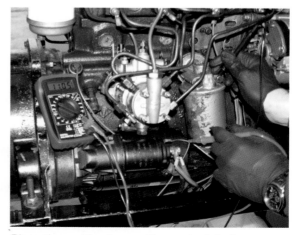

Place the multimeter positive probe into the terminal (make sure the negative is connected to an engine earth) and take a reading when turning on the ignition.

- The multimeter should show a 12V+ reading, indicating that the ignition system is operating correctly. If there is less than 12V or no voltage present, then the ignition system has a fault and needs further investigation.
- If you are confident enough, continue back along this circuit, testing each connection. To do this, you will need to trace the coloured excite wire back along the circuit. Check the following for the presence of voltage in this order: block connector (if applicable); ignition switch; and power feed. By tracing the coloured wire by feel and by eye, you should see any colour changes or connection points that offer an opportunity to retest for voltage. Eventually voltage will show and you will have found the location of the fault.
- If there is voltage present when the ignition switch is in the off position, this indicates a serious electrical issue that requires immediate action. Turn off all isolators and seek professional advice.

ENGINE CHARGING SYSTEMS

Next we will look at the charging system. To aid understanding, this system will be split into two subsystems referred to as 'main' and 'ignition'. *Refer* to the Alternator Connections table later in the section if you need help identifying the connection points on the alternator.

Engine Charging System (Main)

The 'main' refers to the larger cabling, which carries large amounts of power from the alternator to the batteries in order to charge the batteries whilst the engine is running. Faults such as only one battery charging, no charge to the batteries, even though the alternator is indicating it is charging, or intermittent charging indicate a problem in the main charging system. It is worth noting that dirty, loose or poor connections can cause all of the above symptoms, or result in a complete power loss when a system is put under load, for example when trying to start the engine.

The safe charging levels for the alternator are between 13.2V and 14.9V; any higher or lower will result in battery failure/damage in the near future and should not be ignored. It is very common for alternators to fail while still working intermittently, especially if used with a battery management system, as many are designed to override the inter-

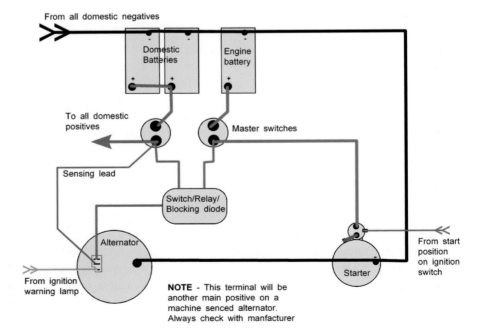

Typical circuit and components in an engine charging system (main) with two battery banks.

From all domestic negatives

Domestic Batteries

Engine battery

To all domestic positives

Master switches

Sensing lead

Switch/Relay/Blocking diode

Alternator

From ignition warning lamp

From start position on ignition switch

Starter

NOTE - This terminal will be another main positive on a machine senced alternator. Always check with manfacturer

nal regulators and therefore the alternator appears to be working. Alternators can also fail but still overcharge the batteries, therefore if left undetected can damage battery banks.

Description of diagram:

- The circuit starts at the alternator.
- Power travels from the alternator to the split charger/voltage relay, which diverts the required charge to the battery banks (only applicable if multiple battery banks are being charged from a single alternator).
- Power is directed through the cables via the battery switch, which has to be on in order to allow the power to reach the batteries and auxiliary components.
- The circuit is completed when the power completes its journey back to the battery.

Circuit testing: To test and trace a fault using an electrical tester, you will need to undertake the following tests to identify the fault location.

Set the multimeter to the Voltage DC setting (V--), choosing a voltage range just higher than you are measuring. Take a reading from the battery to provide a base voltage level for reference (with the engine stopped). The engine should not need to be running to perform these tests, but you may need to start it depending on the components in the system.

- Take the negative (black) probe, connect it to the negative terminal of the battery and leave it in position during the whole test. Where possible, use clips as these will provide a secure connection whilst you test the circuit. If you cannot reach the battery, use any bare metal on the engine/bell housing.
- Place the positive probe on the two connections at the isolator/battery switch and check for the presence of voltage.
- Next, place the positive probe on the split charge connections and check there is voltage present (if the split charger is a diode it will not show

Place the positive probe on the alternator main terminal marked B+ and check for the presence of voltage.

voltage on the alternator side and the engine will need to be started to continue testing the circuit).

- Finally, place the positive probe on the alternator main terminal marked B+ and check for the presence of voltage.
- If voltage is present throughout the circuit, check the negative side of the circuit by choosing the continuity test setting on your multimeter. Place one of the probes on the battery negative terminal and the other probe on the alternator body. The reading should display 0 and an audible signal should sound. If the reading is not 0V, there is a fault on the negative system and further investigation is required.
- The voltage should read considerably higher than the battery base level reading whilst the engine is running, typically between 12.8 and 14.8V. This is because the battery steady state is always lower than when receiving charge from the alternator. If the voltage reading is lower than 12.8V or higher than 14.8V this indicates an alternator fault.
- If the warning light on the panel does not extinguish, this indicates an alternator fault. If the alternator warning light does not illuminate, this indicates an ignition charging system fault, in which case undertake further tests as detailed in the next section.

Components involved in the engine charging system (ignition) required to start charging. Note that failure of a lamp can result in a break in the circuit and therefore charging is interrupted.

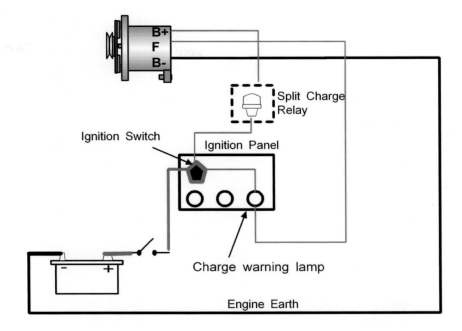

Engine Charging System (Ignition)

The 'ignition' refers to the smaller gauge cabling that excites the alternator and activates the relay needed to direct charge to the correct locations. If the warning light does not come on prior to starting the engine (but with the ignition switch in the first position), or there is no charge from the alternator, this is the system most likely at fault. Please note that loose/corroded terminals will cause all of the above faults and should be checked and cleaned where appropriate.

Description of diagram:

- The circuit starts at any main positive; in this example we start at the isolator switch.
- Power passes through the cables to the ignition switch.
- The ignition switch then transfers the power to the alternator warning light (it may also feed a split charge relay if applicable).
- Power travels through the warning light in the panel to the alternator terminal marked D+ or F (or the small spade connector directly above the two big spade connectors on Lucas style alternators).

Circuit testing: To test and trace a fault on the components within the charging system, using an electrical tester undertake the following tests to identify the fault location. The engine does not need to be running to perform these tests, but the ignition must be on.

Set the multimeter to the Voltage DC setting (V--), choosing a voltage range just higher than you are measuring. Take a reading from the battery to provide a base voltage level for reference (with the engine stopped).

- Take the negative (black) probe, connect it to the negative terminal of the battery and leave it in position during the whole test. Where possible, use clips as these will provide a secure connection whilst you test the circuit. If you cannot reach the battery, use any bare metal on the engine/bell housing.
- Locate the D+ terminal on the alternator. (*Refer* to the Alternator Connections table later in the section if you need help identifying the connection points on the alternator.)

Alternator connections. Shows the typical layout of positive and negative terminals on the back of an alternator.

- Place the positive probe on to the terminal. Once connected, place the multimeter in a position where you can read its display, then turn the ignition key to first position (ignition). The multimeter should display approximately 12V.
- If 12V or more is not present, this indicates an issue with the ignition system. First, check the following: that the fuses are intact; that the block connectors are secure; that a bulb has not blown in the panel (this will break the circuit) and that the isolator switch is on. If the problem is not caused by any of these issues, you will need to seek professional help.
- If you are feeling confident, you may wish to trace the circuit back in order to locate the fault. You can do this by following the same procedure outlined above when checking the starter ignition circuit.
- The components you should check include the following, in this order: block connector (if applicable); warning light; ignition switch; and power feed.

ENGINE HEATING SYSTEM

Next we will show how the heating system works (heater/glow plugs). Again, the system will be split into two subsystems, 'main' and 'ignition'. Due to the complexity of these systems and the number of variants, it is not possible to provide guidance on all testing needed with these circuits. However, using the same tracing techniques outlined in the previous systems, it should be possible to locate any faults not highlighted in the following.

The details below are to help you to identify and understand the different configurations. Having identified these elements, you may be able to troubleshoot. However, as detailed previously these systems are very rarely comparable and therefore cannot be covered within the scope of this book.

Engine Heating System (Main)

The 'main' refers to the larger diameter cabling taking power to the heater plugs themselves. Although in this example the power comes from the

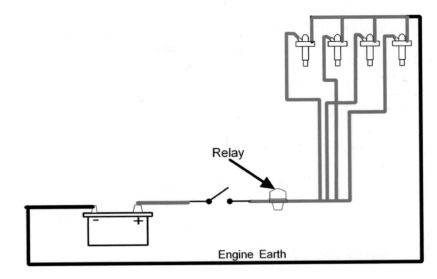

Heating system (main), showing the power to the heater plugs. It may not be evident until winter that this system has stopped working, when the preheat is needed to get the engine started.

Relay

Engine Earth

battery through a relay, the power can come directly from the ignition switch to the heater plugs. Please note once again that any loose or corroded connections will cause faults that will affect starting the engine.

Description of diagram:

- The circuit starts at any main positive; in this example we start at the battery positive.
- Power passes through a relay and travels to the heater plugs.

- Power is returned to the battery via the main earth cable attached to the engine.

Engine Heating System (Ignition)

The 'ignition' refers to the smaller gauge cabling that is used to activate the relay in order to direct the power to the heater plugs. If the warning light works, but the heater plug does not heat, the relay is most probably at fault. Please note that any loose or corroded wiring can result in faults and should be checked and cleaned where appropriate.

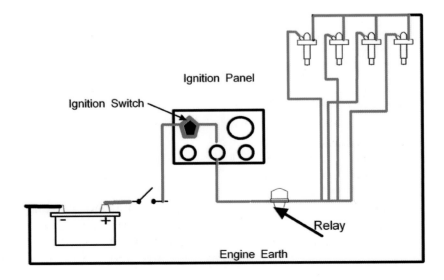

Ignition Panel

Ignition Switch

Heating system (ignition). In this system the heater plugs are fed and controlled via the ignition switch.

Relay

Engine Earth

Description of diagram:

- The circuit can start at any main positive; in this example we start at the isolator switch.
- Power travels through the cabling to the ignition switch, or a button on the ignition panel.
- Power is directed through to the heater plug warning light when the button or ignition switch is activated.
- Power travels through the warning light to the heater plug relay (or it may go directly to the heater plugs).

STOP SOLENOID CIRCUIT

Next, we will look at the stop solenoid system, examining two different subsystems, one called 'live to stop' and the other 'live to run'.

Stop Solenoid Circuit (Live to Stop)

The 'live to stop' system is the type of system that, when engaged, stops the engine. It will usually have a stop button, or an extra position on the key switch (Vetus), or a kill switch. These types of solenoids typically have two wires (red and black).

Description of diagram:

- The circuit can start at any main positive; in this example we start at the battery.
- Power travels through the cabling to the ignition switch/button/switch.
- Power passes to the stop solenoid, which cuts out the engine.
- Power returns to the battery via the main earth.

Circuit testing: To test and trace a fault on the components within the charging system, using an electrical tester undertake the following tests to identify the fault location. The engine does not need to be running to perform these tests, but the ignition may need to be on as some 'live to stop' systems are powered through the ignition system.

Set the multimeter to the Voltage DC setting (V--), choosing a voltage range just higher than you are measuring. Take a reading from the battery to provide a base voltage level for reference (with engine stopped).

- Take the negative (black) probe, connect it to the negative terminal of the battery and leave it in position during the whole test. Where possible, use clips as these will provide a secure connection whilst you test the circuit. If you cannot reach the battery, use any bare metal on the engine/bell housing.

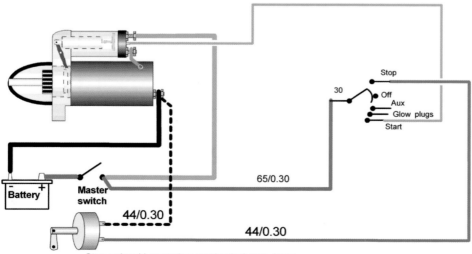

Stop solenoid circuit (live to stop). When the stop button is pressed or the ignition is turned off, the solenoid on the injection pump is engaged (activated) and cuts the fuel to the engine and the engine stops.

Stop solenoid operating mechanical stop lever

- Identify the location of the stop solenoid. When tracing the wires back from the solenoid you will find in close proximity to the solenoid a small block connector or two connecting points.
- Disconnect the block connector and place the positive probe on each terminal end and take a reading with the ignition in the stop position (or with the ignition on and whilst pressing the stop button).
- Voltage should be present at one of the terminals and should reflect the battery voltage. If there is no voltage present at either terminal there may be a fault within the ignition system. (Always take the reading from the loom side of the connector; if taken from the solenoid side no voltage will be present.) If there is voltage present in one terminal (which there should be in a healthy circuit), you will need to perform a check on the negative side of the circuit.
- Set the multimeter to its continuity setting, then place one probe (either one) on the battery negative and one on the terminal that did not have a voltage reading. The audible warning should sound and the reading will indicate 0. If not, this indicates a bad earth connection in the circuit. Trace and check all connections and tighten any loose ones, then retest.

Disconnect the solenoid block connector and place the positive probe on each terminal end and take a reading with the ignition in the stop position.

Stop Solenoid Circuit (live to run)

'Live to run' refers to whether the solenoid is active whilst the engine is in use. In this system the solenoid is active when the engine is running. Please note that if the solenoid fails, the engine will shut down and not restart until the solenoid is replaced. Also, overheating or failing solenoids will cut the engine

Stop solenoid circuit (live to run). The system is live whenever the ignition is on. When the stop button is pressed, or the ignition is turned off, the solenoid is disengaged and cuts the fuel to the engine and the engine stops.

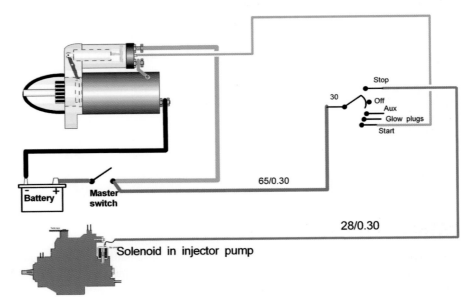

rev, or cut out the engine completely at seemingly random intervals. However, once cooled down, the engine will restart perfectly, leaving no symptoms. It is always worth giving the solenoid a tap, as it can sometimes stick. These solenoids typically have three wires (red, black and white).

Testing on these circuits requires a level of confidence and experience and is therefore not covered in this book. The following details are to familiarize you with the circuits and operations of these systems.

Description of diagram:

- The circuit can start at any main positive; in this example we start at the battery.
- Power travels through the cabling to the ignition switch.
- When the ignition is turned on it sends power to the solenoid. This is designed to hold the solenoid open without damaging the unit.
- As the ignition is turned to the start position, the starter motor begins to turn and high power feed is sent to the solenoid, which makes the solenoid activate.
- When the starter motor is disengaged, the high power stops, but the stop solenoid is held open by the low power provided by the ignition switch.
- When the ignition is turned off the voltage stops and the solenoid closes – shutting the engine down.

The tests and circuits detailed above make up the important electrical circuits that keep the engine operational. As mentioned, these systems, unless newly installed, are unlikely to be easily identified or traced without some challenges. Taking a course or asking an electrical engineer to take you through them is always advisable before attempting any work.

BASIC ELECTRICAL THEORY

This section will provide you with a better working knowledge of electrical circuits. By exploring the theory behind the wires, it details the relationship

TIP: EMPLOYING AN ELECTRICAL ENGINEER

If employing an engineer to undertake electrical work, always use someone who has a current copy of the BMEEA code of practice and who uses the code during his or her working day. He or she should also have gained the BMET qualification.

between voltage, current and resistance and takes you through an exercise to calculate the correct size of cable and the types of protection to install. Although understanding the theory is not essential, it will give you a better understanding of the systems and will be key if you decide to develop your skills further.

A basic electrical system consists of:

- Power source
- Resistive load (light, plug and so on)
- Cabling
- Protection (fuse, circuit breaker).

These basic components are shown in the schematic diagram.

When beginning to explore the world of electricity, it is vital to understand the basics of voltage (V), current (I) and resistance (Ω) and how they relate to each other within an electrical circuit. At first, these concepts can be difficult to understand because we cannot 'see' them. No one can see the current flowing through a wire or the voltage of a battery sitting on a table. In order to determine voltage or current, we must rely on measurement tools such as a multimeter.

The accuracy and way these instruments measure values are key to ensuring that the information obtained is reliable. Before undertaking any measurements, always ensure that your meter is working correctly. This can be done in several ways; the most common is by using a 'proving' unit. As most

Basic electrical circuit with a lamp, switch and protection.

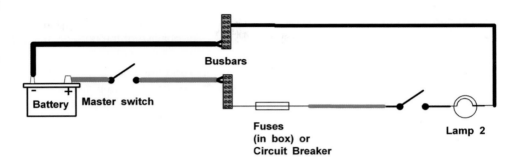

of you will not have such a device, try measuring something with a known value. For example, engineers' measuring equipment will be regularly calibrated, so you could ask to compare a particular measurement on your multimeter to the reading on theirs and note any variation. This would give a figure that would allow you to correct the reading on your meter.

The next few paragraphs provide a basic understanding of voltage, current and resistance, and how they interact and work with each other.

All circuits contain these elements:

- **Voltage:** Something has to make the current flow in any wire and it is known as voltage. Voltage is measured in volts and has the symbol V. (Think of voltage as a characteristic of electricity defining 'how energetic the electricity is'.)
- **Current:** The intensity of current flow, or the amount of current that is flowing for a given time. Current is measured in amperes and has the symbol I. (Current or amps can be thought of as the amount of strength the electricity has, relative to the amount of space it takes up in the wire.)
- **Resistance:** The naturally occurring property of a material to resist the flow of current, depending upon its cross-sectional area and length. Resistance is measured in ohms and has the symbol Ω. (Resistance can be thought of as friction.)
- **Power:** Usually a measure of how much a machine generates or an appliance consumes. It is derived from the relationship between voltage, current and resistance. Power is measured in watts and has the symbol W.
- **Voltage drop:** The drop in voltage measurement caused by resistance.

The relationship between voltage, current and resistance means that any of these components can be identified as long as two of the quantities are known:

- Voltage can be found by multiplying the current (in amps) by the resistance (in ohms): $V = I\,R$.
- Current can be found by dividing voltage by the resistance: $I = V / R$.
- The resistance can be calculated by dividing the voltage by the current: $R = V / I$.

This simple formula is called Ohm's law, and is probably the most important element in electrical theory.

- If the voltage and current are known we can also work out the power, because the power (in watts) is equal to the voltage (in volts) multiplied by the current (in amps): $P = V\,I$.

And, much like Ohm's law, if two of the values are know the formula can be used to identify the missing one:

- to identify the voltage if the power and current is known: $V = P / I$
- to identify the current if the power and voltage are known: $I = P / V$.

Generating Current AC/DC

Alternating current (AC): If a coil of wire is rotating within a magnet field, this induces a magnetic flux within the wire that becomes flowing electricity. As the wire rotates, the strength of the flux will rise and fall, causing an oscillating effect. This is known as an alternating current and follows a sinusoidal wave. Generators, alternators, wind generators and so on will all produce AC and require a rotating section within the equipment in order to generate the current. AC is what every household in the world relies on to deliver power for everyday needs. The shore power and inverter power on a vessel will be AC and usually supplied at 230V.

Direct current (DC): Direct current does not vary – that is, it is static and is usually delivered by batteries, solar cells or dynamos. However, batteries once discharged have to be topped up. This requires an alternator to generate the current and then convert it to DC in order for the cells in the batteries to accept the charge. The 12V/24V systems on a boat will be DC systems, with the battery charger or the alternator supplying the top-up power.

NOTE: The 12/24V alternator fitted to your engine is an AC machine that incorporates a device (a rectifier) to convert the output to DC at the terminals.

In the diagram, we use the common approach, which states that DC current will flow from the red or positive of the battery, its path will travel via the circuit breaker and lamp and return to the battery along the black or negative wire. Therefore, in this circuit the current flow is from the positive to the negative of the power supply or battery.

A circuit requires that the power flows full circle from the power source through components and back to the power source. Any breaks in this circuit and there will be no power flow and therefore all the components will stop functioning.

When there is a break in the circuit, this is referred to as an open circuit, that is, the circuit is not closed and therefore the power cannot flow. This can be caused by loose connections, poor battery terminals, or breaks in wires.

However, the biggest cause of major issues is when the circuit suffers a short circuit. This is usually due to an overload condition, or when incorrect wiring has been used and the 'normal' route around the circuit has been short-circuited. This condition will usually result in wiring loom meltdown, component failure or fire. It can happen anywhere in the circuit and is the reason that protection of the circuit through the use of fuses and circuit breakers is essential. Without protection, the short can potentially affect the whole circuit, that is, a meltdown can travel unheeded until the circuit is broken.

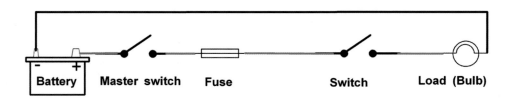

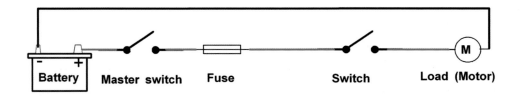

Electrical circuit with a simple load and a motor.

By placing an electrical tester on the circuit at different locations and reading amps, volts or ohms, it should be possible to determine the value of these components. This will enable you to determine the protection or cable size required. In addition, you can undertake a continuity test. This setting allows you to connect the meter to two separate parts of a circuit and check that there are no breaks in the circuit. It is also valuable when tracing wires and cables and checking circuits. The continuity test usually has an audible alarm function and is denoted on the multimeter with one of the following three symbols: ∞ ¥ ⅄.

The circuit breaker shown in this circuit can be replaced with a fuse, as both provide protection functions. When a circuit experiences a short or other electrical issue, the protection will operate to indicate an issue and also to protect the system from high currents that can flow when these faults occur. In many circuits, there will also be an isolator, which is there to allow the system to be isolated.

Isolation and protection provide very different functions. One has to break the current during a fault condition, whilst the other isolates a system in its normal state. Each is therefore rated very differently.

Overcurrent Protection

A circuit breaker will have a continuous rating and a peak rating. It will also detail how long it takes to react and will therefore indicate how long your system will experience the fault condition before the protection operates. With 12V protection, ensure that the breaker is rated for DC rather than AC, as use of an AC breaker on a DC system will result in contacts fusing or welding.

A fuse has a max amps rating and will be provided with information on fuse burn time. Some fuses have a slow burn, which means that they can handle higher currents for a longer period of time before they fail. A fuse consists of a piece of wire that is exactly the right length and thickness, so that when a current rises to a specific value it will overheat and break. When a fault occurs and the current rises, the fuse should 'blow' rather than any other part of the circuit overheating.

With overcurrent protection, it is essential that it is able to withstand a higher current than the appliances to which it is connected, but a lower current than the cables that connect them. This ensures that in the event of a fault, the protection operates before the cable is damaged.

The resistive load can be any passive device, from lights, sockets, wire and so on. However, a device can be capacitive or inductive, depending upon how it uses or holds current. Typically, if an electrical motor is used within the circuit this will be an inductive load and therefore requires additional consideration when thinking about cable sizing and the systems to protect it.

Circuit Design

With any circuit, selecting the correct cable size is one of the most important elements to ensure safe operation. If the cable size is incorrect a circuit may not function correctly, or will result in faults developing in the future. Cable sizing is dictated mainly by voltage drop, which is the amount of losses that will occur when voltage is travelling through a cable, and is related to the length and diameter of the copper conductor within the cable. These losses are related to heat, current and voltage, and magnetism; they are not covered within this book.

With current regulation, there are two figures that are commonly referred to as 3 per cent and 10 per cent voltage drop, which is the minimum safety margin that should be used when sizing cables. The 3 per cent rule applies to all 'critical' circuits, such as bilge pumps, navigation and radio equipment. All other circuits should function with a 10 per cent voltage drop.

Maximum Amount of Voltage Drop for 12V and 24V Circuits

	3%	10%
12V	0.36	1.2
24V	0.72	2.4

Cable Sizing Example

All DC equipment should be capable of functioning within a voltage range of 75–133 per cent of nominal voltage at the battery terminals, for example:

- for a 12V system, equipment should be able to cope with 9–16V
- for a 24V system, equipment should be able to cope with 18–32V
- for a 48V system, equipment should be able to cope with 36–64V.

When considering extra-low DC voltage, a different calculation is used, as the voltage drop on load has different characteristics; these situations are not covered in the book.

The following formula is used to identify the correct cable sizing:

$$E = \frac{0.0164 \times I \times L}{S}$$

Where:

S   is the conductor cross-sectional area (CSA), in square millimetres

I   Is the load current, in amperes

L   is the length, in meters, of conductor from the positive power source to the electrical device and back to the negative source connection

E   is the voltage drop.

Using a simple schematic, we can work through a calculation (*see* diagram).

In the first example:

- the cables will have a CSA 0.75mm^2 (S)
- the load is 40amp (I)
- the cable length is 20m (L)
- $0.0164 \times 40 \times 20 = 13.12$ and this is divided by the CSA of 0.75(S) = 17.49V.

So in this calculation we have a voltage drop of 17.49V, which is greater than the 12V battery supply and shows that the cable is far too small for the cable run.

Assuming that by increasing the cable size we will resolve the issue, we can try to recalculate using 2.5mm^2 cable, but this still gives a voltage drop of 5.25V, which is still too high (max = 1.2V).

So to calculate the correct cable size, we need to rearrange the formula to identify the value of S (the cable size), given the set parameters and the maximum voltage drop we can allow, that is, 10 per cent of 12V:

- S = the cable CSA
- 1.2 = maximum voltage drop (10 per cent of 12V) (E)
- 13.12 = the coefficient (L * I * 0.0164)
- S = 13.12/1.2 = 10.93   S = (L*I*0.0164)/E.

Having completed the calculation, the figure we have is 10.93 and this is the minimum cable CSA that we can use in this circuit. As 10.93 is more than 10mm^2, always select the next cable size up, which in this case will be 16mm^2. There are calculators on the Internet that you can use to work out the voltage drop by the touch of a few buttons.

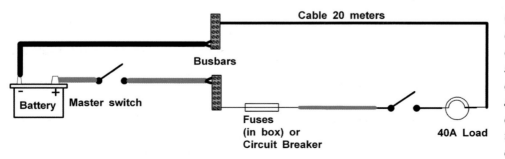

Cable calculation circuit, showing the components and values detailed in the cable calculation.

If your boat was constructed within the last few years, it is covered by the Recreational Craft Directive. The directive makes provision for two manuals: one is retained by the boat builder; and the other is for the owner. Both of these manuals are compiled by the boat builder. Within the manual retained by the boat builder should be all of the voltage-drop calculations for each circuit installed within the boat. Unfortunately, these are often missing.

Hopefully the above sections should have opened the doors to calculating and checking that the correct cabling is used in circuit design or repair.

ALTERNATORS

Vessels are becoming more complex and many now require significant direct current (DC) charging systems. These are usually 12V or 24V systems. An alternator is just like a generator, in that it transfers the rotational movement from the engine via the fan belt and converts it to electrical power. The alternator generates an alternating current (AC), but this is then converted to 12/24V DC. The current generated is used to the charge the batteries and provide power to any other ancillaries that require power.

Alternator Design

An alternator is used to recharge batteries, but because it produces AC this has to be converted to DC in order to provide a charge current that a battery can accept. An alternator designed for this purpose has three coils of wire arranged in a fixed position (known as the stator) and an electro-magnet that rotates between them (known as the rotor).

The alternator output is governed by three factors:

- The turns of the stator winding
- The speed of the rotor
- The strength of the magnetic field within the rotor.

The stator windings are determined by the manufacturer's design. The speed at which the rotor turns is governed by the engine's rpm and the pulley sizes used, but the speed will be always changing. The only thing that can be regulated is the magnetic field

within the rotor. To achieve this, a regulator is used inside the alternator.

Most modern alternators contain a regulator for voltage regulation, which changes the strength of the magnetic field within the rotor and thus controls the output. In addition, there will be a rectifier to convert the AC output of the alternator into DC. The rectifier is made up of a number of diodes that are used to change the AC output into a DC output.

Alternators are designed to produce a higher voltage for battery charging. A higher voltage is required to overcome the internal resistance of the battery and since boat batteries are usually rated at 12 or 24V, the alternator will have to produce around 14–15 or 28–30V.

The voltage output is regulated by the voltage regulator controlling the magnetic field within the rotor. This is controlled by adjusting the pulsing of the current within the rotor.

Alternator Operation

The standard specifications for most starting alternators, that is, ones that solely charge the starter battery, range from 35 to 70amp. However, if it

Typical alternator connections. This Beta engine has a starter alternator on the right and the larger domestic alternator on the left; both are driven by the large crank-shaft pulley.

is a dual alternator, that is, charging the starter and domestics, they are usually in the range of 50–90amp. For domestic alternators, that is, ones that charge only the domestic batteries, these are in the range 70–175amp. These higher output alternators will normally require a multi-belt drive (flatbelt) to avoid the belt slipping and wearing excessively. You need to identify the type and ratings of the alternator/s installed on your vessel and make a note of them.

If your vessel has a single alternator, you will find that the usual arrangement is to have a split-charge system fitted to the battery supply. This splits the charge between batteries and ensures that both banks are electrically separate, so that if one is drained it does not affect the other.

If your vessel has dual alternators, each system should be wired separately so that if one system fails, for example the domestic alternator fails and the batteries are drained, the engine battery is unaffected and the engine can be started.

TIP: A 'GET YOU HOME' SOLUTION

If you have a dual system and one alternator fails, use a single jump lead to connect the positive of one battery bank to the positive of the other battery bank. This will allow the single alternator to charge both banks of batteries while the engine is operating. However, the jump lead must be removed when the engine is switched off, or it will result in the engine battery being discharged along with the domestic batteries should you use too much power for the domestic systems.

Alternator Pulleys

It is a common misconception that an alternator rating indicates the amount of power or current that an alternator will deliver; unfortunately, the speed of the alternator, its operational temperature and the load it is supplying can all affect alternator output. The most important factor in achieving the full capacity of the alternator is the sizing of the pulley, as this has to be closely matched to the normal engine speed. On a vehicle, the engine will generally be running at max rpm and therefore the pulley is matched to achieve full alternator output at this maximum rpm. However, on an inland waterways boat, where the typical speed is less than 4mph, achieving the maximum charging rate can mean calculating the correct sized pulleys to suit the new engine operation.

Newer engines supplied for marine application will already have been maximized by the supplier and some older engines that have been correctly marinized will also have been modified to ensure correct and effective charging. However, considering the range of second-hand, DIY and marinized vehicle engines, it is always worth doing the checks to ensure that you are getting the optimal charge from your system.

First, you need to calculate the rpm of the alternator and to do this you need the measurement of the crank-shaft pulley and the alternator pulley size. Divide the crank pulley size by the alternator pulley size to get the ratio:

- **Example:** If you have a 4in crank pulley and a 2in alternator pulley: 4/2=2.

In this example we have calculated a two to one ratio; multiply the engine rpm by this value to calculate the speed of your alternator rpm. In this example we use an engine rpm of 1,000, multiplied by the ratio. This indicates that the alternator is operating at 2,000rpm. When you calculate this for your engine, make sure to use your normal engine running rpm in the calculation and aim to have a ratio higher than 3:1 to ensure that the alternator is producing enough power at low engine revs.

Single Alternator

The majority of older engines and those below 30hp will usually be supplied with a single alternator, which has to supply charge to the engine battery and the domestics. With systems where only one alternator is employed it is important to ensure that the starter battery and domestic battery bank are isolated (that is, not connected), otherwise you run the risk of depleting all the batteries if your domestic demand is too high, resulting in not being able to start the engine.

The battery banks need to be wired so that they are isolated from one another. This is achieved with the use of a split-charge diode, or voltage-sensing relay. The split-charge diode acts like a one-way valve, as it only allows voltage to flow one way; it will allow charge to be split between two banks, but keeps them isolated from one another. The split-charge relay will automatically connect the battery banks as soon as the alternator begins to charge and then stops and isolates the battery bank when the alternator stops charging.

An alternative is to fit a voltage-sensing relay that is controlled by the level of current being produced by the alternator. The higher the current, the lower the voltage, as the alternator depresses the voltage when delivering high currents to batteries. By monitoring the voltage and using this to control the internal relay, it ensures that the battery bank requiring the highest charge will receive the bulk of the output.

Twin Alternators

On most modern engines or those above 30hp, the engine is supplied with both a starter alternator and a domestic alternator. These are usually different specifications, so the starter alternator typically will be from 30 to 70amp, whereas the domestic alternator will vary from 50 to 175amp. This dual-alternator arrangement provides separate charging systems and as the starter alternator is only charging one battery, this will be recharged relatively quickly, so for most of the time the engine alternator is operating on light duty.

Alternator Output

The alternator characteristics dictate how batteries are recharged. A typical charging curve of an alternator is shown in the diagram.

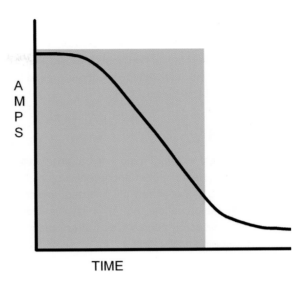

Alternator charging curve, showing the amount of charge that is delivered by an alternator over a period of time.

Within a few minutes of the alternator starting to charge a discharged battery, it will be delivering its maximum output as dictated on the alternator plate or detailed in the engine manual. The output from the alternator remains at a maximum for a period of time until the battery begins to charge. The output then gradually falls, until only a 'float charge' is being delivered.

Unfortunately, as it is impossible to identify how long the float charge will be supplied, it is also not possible to work out an average charge to use in a calculation. However, using the alternator charging curve, we can assume that the average will be about half the alternator output, which generally works out at approximately three hours. The time represented by the shading in the diagram will alter depending upon alternator output and battery bank size. If you have an ammeter (and ideally a voltmeter), you will be able to monitor the alternator and batteries more accurately.

This information is used below in the section on batteries, as it feeds into the calculations on the length of time it takes to recharge a battery.

Alternator Connections

Inspecting the rear side of the alternator, you will usually find a number of terminals marked D+, W and B+. These indicate what connections should be made (*see* the table below).

BATTERIES

There are many different types of batteries available and each varies in cost and specification. The following is a quick overview of the types, arrangements and maintenance, but it is always worth doing some research yourself, as battery technology is developing all the time.

Nickel-Iron and Other Non-Lead Based Batteries

These types of batteries are not too expensive, but have a very long life and because it is possible to change out faulty cells it makes them ideal for domestic systems. However, they do require a dedicated charging system and it is important to ensure that you have access to the specialist services needed to maintain and support your installation.

Engine-Starting 'Cranking' Batteries

These batteries are usually lead acid and constructed with a number of thin plates, so when there are high current flows the outer surfaces suffer as little stress

Alternator Connections

Marking	Purpose/Comment
+B, +BAT, BAT+, = 9mm male blade	Main positive output connection going *to the battery* (master switch), usually thick red or brown cable.
D+	As long as there is no B or B+ this is the main positive output connection.
D+	If there is a B or B+ this is an auxiliary output connection (for external control gear, or split-charge relays, or a warning light wire) and is usually brown with a yellow stripe (a small spade terminal situated directly above the two large spade terminals on a Lucas-style alternator*).
–B, –E, Gnd, –Bat,	Main negative connection. This either runs to the engine block, or the battery negative terminal.
6mm male blade F DF	Warning lamp connection.
61, Ind, L, Aux	Auxiliary output for external control gear, or split-charge relays and so on.
W, AC	Phase tap for alternator-sensed rev-counter (Tacho) and is usually blue.

*Lucas alternators use a 3-pin plug; the two bigger pins are usually equivalent to B+ and the smaller one is D+, although there will be a difference in wire colours. The main positive cables will be brown and may be split into two smaller gauge brown wires to fit both pins.

NOTE: There may be a single unmarked 6mm blade set into the case. This is for a radio suppressor that may or may not be fitted to the alternator.

as possible due to a higher surface area and consequently they shed minimal amounts of their material. They are designed to supply cranking power – high current demand in short bursts – without suffering damage.

Deep Cycle/Domestic Batteries

These batteries are designed to provide low current flows for long periods, so are ideal for domestic batteries. They are usually constructed with thick plates contained in a fibreglass pocket that holds any material that the plate sheds. These batteries can be discharged and will recover without significant losses. However, the deeper the discharge, the shorter the life of any battery.

Traction or Semi-Traction Batteries

These are usually constructed with plates formed as tubes or rods. They are expensive, but have long lives and are installed with a dedicated charging system to ensure that they are maintained correctly. Usually supplied as 2V cells, they can be built up to supply any amount of current consumption and due to their size can be easily fitted to small spaces, which makes them ideal for boat applications. Typical life is ten years, but many are still in operation twenty-five years later when well maintained. They are not suitable for engine starting.

Maintenance-Free or Sealed for Life

These batteries offer low self-discharge, so provide advantages when a boat is left for periods of time when not in use, or for the whole winter period. As their name suggests, there is no maintenance apart from keeping the tops clean and the terminals dressed. Unfortunately, they have been known to explode if subject to very high rates of charge, or where gases produced due to discharge cannot be vented. When totally discharged (for example, due to leaving a light on), they may require specialist charging to recover them.

Gel Batteries

The acid is in a form of jelly, so that it cannot leak out. These are relatively expensive, but have low self-discharge rates, so hold their charge well. They cannot accept high charge rates, so once discharged can take a long time to recharge, but the main benefit is that they do not spill if tipped. For this reason, they are ideal for ocean-going yachts; however, on the inland waterways they do not offer much benefit.

Dual-Purpose Batteries

These batteries are the most common batteries installed for domestic applications. However, they are simply a modified starter battery and therefore prolonged usage may result in early failure – if they are regularly discharged beyond 40 per cent, it is unlikely that they will recover. As a domestic battery, they offer a cheap alternative, which is why there are so many installed in boats.

If you only use the vessel for holidays, using dual-purpose batteries may suffice, as the chances that the batteries will be discharged would be unusual. However, if you live on-board the boat, or use it for extended periods, you should invest in true domestic batteries.

The starter battery should also be chosen based on the needs of your engine. In most cases, the starter battery will be lead acid and the domestic bank should be matched to your needs. Battery selection and the amount required should always be based on your personal usage.

Battery Banks

Several batteries can be linked together to give almost any combination of capacity and voltages. It is important to identify the battery capacity, arrangement and voltage before you consider upgrading, making changes and carrying out maintenance. When connecting the batteries in series or parallel, different configurations can be achieved. Some of the more common examples are shown below.

In series: Two 6V batteries connected positive+ to negative– to give 12V, but with the same capacity as a single battery.

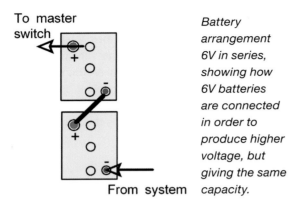

To master switch

From system

Battery arrangement 6V in series, showing how 6V batteries are connected in order to produce higher voltage, but giving the same capacity.

In parallel: Two 12V batteries connected positive+ to positive+ and negative– to negative– to give twice as much storage capacity, but only at 12V (probably the most common method).

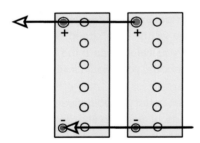

Battery arrangement 12V parallel, showing how 12V batteries can be connected in parallel to increase the capacity (double the amps), but remain at 12V.

You can keep adding batteries, connecting positive+ to positive+ and negative– to negative–, to achieve the capacity you require.

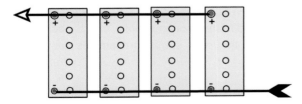

More batteries can be added in parallel to increase current capacity.

In series: Two 12V batteries connected positive+ to negative– to give 24V, but with the same current capacity as a single battery. This is a common arrangement for 24V systems and can be used for starting and domestic applications.

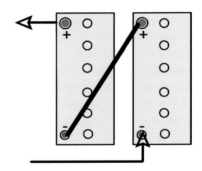

Battery arrangement 12V in series, showing how 12V batteries can be connected in order to increase voltage to 24V, but the capacity (amps) output remains the same.

In series: Four 6V batteries connected positive+ to negative– to give 24V, but with the same capacity as a single battery.

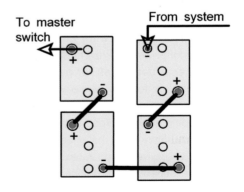

Battery arrangement of 6V batteries in series, which will increase voltage to 24V with the same capacity.

Locating the Batteries

If you are having a new boat built, modifying one, or doing an extensive refit, think carefully about the position of the batteries and consider the following points:

- Ensure that you have easy access to all the batteries to check, top-up, service and replace them.
- Locate the batteries where the gases (hydrogen and oxygen) can escape to the open air. Consider the damage that an exploding battery could do to the surrounding area and the effect of spraying acid about.
- Site the batteries as close to high-current equipment as is reasonably possible – or locate the equipment close to the batteries, as this minimizes long runs of large, expensive cables. Typical equipment would be starters, alternators, inverters and electric fridges.
- Try to keep them as vibration-free as possible; mounting on rubber mats reduces the risk of sparks and damage to the battery.
- Keep them as cool as possible and allow gases to escape by providing vents in the battery box; this is a BSS requirement.
- Ensure that the batteries are secured using straps (BSS requirement); this stops them jumping about and making contact with the hull.
- Fit some type of insulating cover to the battery box. This will reduce the risk of metal objects dropping on to the battery and causing sparks or an explosion. Never store anything other than batteries in the battery compartment (BSS requirement).

Battery Care and Service

It is important to appreciate that battery charging can affect battery life. Any batteries that are left in a state of partial discharge will suffer a permanent deterioration in their plate material, which can never be recovered and will eventually result in their failure. Caring for your batteries will reap benefits, both in terms of reliability and life. Battery care is easy if done often and following some simple rules:

- Keep the battery top clean and dry and the terminals clean, corrosion-free and treated with dressing/petroleum jelly. This will prevent discharge through the damp acid/dirt on its top and stop corrosion build-up on the terminals.
- Keep the battery topped up with distilled or demineralized water. Never overfill, as this can result in excess acid leaking out of the battery, which will corrode any metalwork in the vicinity. The level on the battery will usually be marked with a level line on the side (you may need to shine a light down through the top of the case to see the acid level). Alternatively, when looking down through the filler cap, you will find the level marked with a bar or half-moon piece of plastic.
- Inspect the ends and sides of the battery for signs of bulging. When this appears, it indicates that the battery is ready for replacement, or may have been damaged due to overcharging.
- Keep the batteries as fully charged as possible – even if it means taking them home during the winter for servicing and charging. Keeping the batteries in a good state of repair will ensure that the longest possible battery life is achieved.

Testing Batteries

There are two methods of testing batteries that can be performed without the need to remove the battery and take it to a specialist. Each method is discussed below, along with how to perform the tests.

Measuring relative density: Lead acid battery electrolyte has a relative density of between 1.150 and 1.250, depending upon the state of charge and temperature. Measuring the relative density of the acid in each cell provides an indication of the state of the battery and gives the most reliable results. This is achieved using a hydrometer, which can be bought from most auto shops, although this test can only be performed on batteries where the electrolyte can be accessed. Unfortunately, gel batteries can only be tested using specialist equipment.

Hydrometer.

Hydrometer: The best battery hydrometers are made of glass or plastic tube with a rubber bulb at one end and a rubber tube at the other. A float inside the glass tube allows the relative density of the electrolyte to be measured when a sample is drawn up into the glass tube.

One side of the float is normally graduated between 1.100 and 1.300, whilst the other is usually colour coded in bands to show discharged, half charged and fully charged. It is recommended to use the numbered graduations to get the most accurate measurement.

TIP: BATTERY HANDLING

Take care when working with battery acid – it will burn both your skin and your clothes. Also remove watches and be aware of any zips and such like, as any metal to metal contact can produce a spark and ignite gases in the vicinity.

Testing a Battery Using a Hydrometer

Ensure the cells' electrolytes are at the correct level, otherwise there will not be enough fluid to draw into the hydrometer. If the cells are low and need topping up with distilled water, it is important to place the battery on charge for a while to ensure the water is

mixed with the electrolytes before proceeding with the test. To obtain the most accurate readings, the test should be done at an ambient temperature of 15°C.

Wearing suitable protective clothing (rubber apron and goggles) and taking great care, draw a sample of the electrolyte from each cell in turn. Note both the reading and colour for each cell individually. The readings taken indicate the state of charge of each cell. A faulty cell will have a lower state of charge and this dictates the overall battery condition. The electrolyte should be clear, like water. If it is brown or grey, it indicates that the plates in that particular cell are shedding material and the battery is therefore at or nearing the end of its life. If the readings in the cells differ by more than 0.050 (many sources advise values between 0.03 and 0.025) the battery should be replaced and further tests, measurements or calculations do not need to be undertaken.

A hydrometer taking a reading from a cell in a battery. Most batteries now have a display to give an indication of the battery condition (usually red/amber/green).

Once you have all the readings from the cells in a battery, average them and check against the table to identify the charge state of the battery, for given relative densities. Some batteries have a display on them and the colour indicates the battery state. If you are unsure about the battery condition, it is always worth topping it up, charging and retesting it.

Battery Charge State (from Hydrometer readings)

Charge State	Average Cell Charge
Fully charged (green)	1.280
Half charged (orange)	1.200
Fully discharged (red)	1.150

Testing a Battery Using a Voltmeter

The voltmeter test is not as accurate as using the hydrometer and there is ample scope for false readings. However, it is useful if you need to make a quick check on a battery's state of charge and condition.

Before undertaking the test it is important to remove the surface charge; this is a charge that builds up on the surface of the battery plates as soon as a charge is applied. It will give a false reading, because it can be present even if the rest of the plate is discharged. This is because the speed at which a lead acid battery is charged is regulated by the rate at which the chemical reactions take place inside the battery. When charging starts, the chemical reaction is only present on the surface of the plates and it takes time before it starts penetrating deeper into the plates. Discharging a battery surface depends upon which type of battery you are testing:

- If this is the engine starting battery, spinning the engine on the starter for a count of twenty will remove any surface charge that has built up.
- For domestic batteries, turn on all the lights, fans and pumps for approximately 10–15sec per battery in the battery bank. Sometimes more accurate results can be achieved by fully charging the battery bank, switching off the battery

master switch, then leaving for several days before testing, as the charge has a chance to penetrate deeper into the plates and provide a truer picture of the battery state.

Having removed the surface charge, connect the voltmeter across the battery and take a reading using a multimeter. *See* the table for an explanation of the readings.

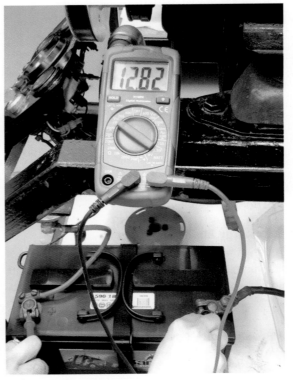

To check the battery voltage, connect the multimeter positive and negative probes to the corresponding terminals on the battery and take a reading.

Multimeter Readings

Very discharged	Less than 12.2V
Partly charged	12.2V–12.5V
Fully charged	More than 12.5V

This method only provides an indication, not a definitive measurement about the battery state. If you have a badly sulphated battery that is otherwise in good condition, this would show as being fully charged but then discharge in minutes. If you have solar panels or wind turbines on-board, they can also provide false readings, so always disconnect or isolate these before taking battery measurements.

Calculating Battery Capacity

The engine starter battery required specifications can be identified by checking with the engine maker/marinizer/dealer. As a general guidance for typical inland waterways craft, a 110amp hours battery will usually be sufficient (equivalent to a transit van starter battery).

The first step to understanding your domestic electrical needs is to undertake a power or energy audit. This will give you an idea about your electrical demands and will enable you to calculate the battery capacity required to match it. From here, we can then calculate the time that it will take for the battery to recharge.

The Energy Audit

The best way to approach an audit is to make a list of every electrical component on the boat and its normal consumption. For situations where the electrical component only quotes wattage on the plate, you will need to calculate the current consumption. To do this, for all components fed by 12/24V directly we use the system voltage and divide the wattage by 12V or 24V respectively. In addition, we need to account for the losses of inverter loads. Using the inverter wattage, divide by 10V or 20V respectively instead of 12 or 24. This takes into account that an inverter is only 80 per cent efficient and therefore the full voltage capacity is never experienced (80 per cent of 12V = 10V; 80 per cent of 24V=20V).

Once you have listed all the components, you will need to identify the maximum amount of time that they are used for within a 24hr period and use the 'worst-case scenario' data. Then multiply the amps by the hours used and total all the results as shown in the table.

Having completed an energy audit, you will have an indication about how much power your batteries need to supply, but first we have to convert it into battery capacity, and this is not as simple as it may appear. Once a battery starts being used its capacity starts to reduce, therefore we have to take into account this deterioration in our calculations.

Actual Battery Capacity

As soon as a battery is used its 'achievable capacity' will begin to decrease. The type of battery, qual-

A Power Audit – Low Consumption

Appliance	Current Rating (Amps)	Time Used (Hours)	Amp Hours (Ah) Required
Fridge	Gas fridge, no power use		
Lights (LED)	4*0.1	6	2.4
Water Pump	8	2	16
Toilet	Manual cassette, no power use		
Central Heating	Stove, no power use		
Radio /CD	<1	5	5
		Total	23.4Ah

A Power Audit – Medium Consumption

Appliance	Current Rating (Amps)	Time Used (Hours)	Amp Hours Required
Fridge	2	18	50.4
Lights	4*1.8	6	43
Water Pump	8	2	16
Toilet	15	1	15
Central Heating	1.5	6	9
TV	3	6	21
		Total	154.4Ah

ity and construction will all impact on the rate of deterioration. The alternator charging efficiency and usage will also impact on the 'retained' capacity of the battery. For this reason, we assume only 80 per cent of battery power is available to us.

As a rule of thumb, if a domestic battery is discharged to below half-charged (50 per cent), a percentage of the capacity will be lost, so the more often and deeper a battery is discharged, the shorter its life. This is why it is important to ensure that you have a good charging regime and match your batteries to your consumption.

We therefore have to take into account that we cannot discharge a battery beyond 50 per cent when we calculate actual battery capacity. Using the base of 80 per cent charge within the battery, we must deduct the 50 per cent and this leaves us with 30 per cent of actual capacity. In reality, this means that we can only use 30 per cent of what is written on the batteries during normal operation.

Therefore using the calculated requirement for the domestic bank of 154.4Ah (from our power audit) and taking into account that we can only actively use 30 per cent of the battery capacity we would actually require:

- 154.4Ah/30 × 100 = 514Ah.

This indicates that a battery bank of 5 × 110Ah batteries is needed in order to deliver the required power.

There are only a few things that will influence the values used in the calculation. For example, if you have a battery-management system installed or the regulator upgraded, this can increase the charge achieved to 95 per cent of fully charged.

Having identified the required battery capacity in order to provide adequate power, we also need to identify how long it will take to recharge the batteries. This will depend upon the alternator operation. As discussed in the previous section, we know that in general the actual alternator output will be about 50 per cent of its rated output and it will take approximately 2–3hr to achieve this value. We use this information to calculate how long it will take to recharge a battery.

Recharging Batteries

Guidance on recharging batteries is given in terms of a percentage of the battery capacity and figures quoted range from 10 to 40 per cent. This guidance takes into account temperature and battery condition and that the higher the charge rate, the more the battery life is shortened. However, on a boat the batteries are rarely provided with a continual charge and therefore a 20 per cent figure is more practical.

Using the example we calculated above, that is, a 154Ah load, and assuming that we need to charge at 30 per cent more capacity, we would need to put back 130 per cent of this figure, so $154 \times 1.3 = 200$. This means that during each charging period we need to put back in 200Ah. If the alternator is rated at 70amp, using the alternator charging curve we know it will deliver about half its rated value, that is, 35amp over two to three hours, depending upon the pulley size and engine speed. Therefore in order to identify how long it would take to charge the battery we take the calculated amps and divide by the charge:

- 70amp alternator: $200 \div 35 = 5.7$.

From this calculation we can identify that we would need to charge the battery bank for approximately 6hr in order to fully charge it, which is why bigger alternators are installed on the domestic systems:

- with a 90amp alternator: $200 \div 45 = 4.4$hr
- with a 140amp alternator: $200 \div 70 = 2.8$hr.

The above calculation does not take into account the effect of the pulley and alternator rpm. However, it should give a good guide as to what battery capacity is required and how long you will need to charge the batteries in order to maintain the battery life.

Battery Fault-Finding

A battery that is under load, that is, not isolated, should always show as having a residual charge greater than 10V. If it does not, check the following:

- state of charge
- cell condition (*see* hydrometer tests).

The following are all problem signs:

- If the top of the battery is covered in white splodges/dust, or is wet, the alternator voltage regulator is probably faulty and causing the alternator to overcharge.
- If the battery smells of rotten eggs or is hot to the touch, either there may be a cell/cells that are failing, or the alternator voltage regulator is faulty.
- If the ends of a battery are 'bowing' (distorting the casing), this indicates that the battery is nearing the end of its life.
- If all the cells need constant topping up, it indicates that the charging voltage is too high or that the ambient temperature is consistently too hot.
- If a single cell or two needs constant topping up, it indicates a faulty cell, so the battery will need changing.
- If the voltage across the battery when on charge and after about an hour is less than 13.6V, this indicates either a slipping alternator belt or a faulty alternator. (The longer a battery is on charge, the higher the charging voltage. Expect up to 14.3V on a normal regulator and more with an advanced one.)
- Brown/grey/blackish electrolyte indicates a faulty cell.

TIP: BATTERY SAFETY

If you are unfortunate enough to experience a battery explosion, the acid can be neutralized by using bicarbonate of soda, which should be liberally spread over the entire area to prevent corrosion and acid deterioration.

CHAPTER 5

PROPULSION SYSTEMS

INTRODUCTION

Propulsion is the term used to describe how the energy from the engine is used to propel the vessel through the water. It transmits power from the engine through a series of components to the propeller and thus creates drive. Hybrid drive systems and electric drive systems are not covered in this book.

The propulsion system section covers the output from the engine, the gearbox and the adjoining prop-shaft couplings through to the prop-shaft, stern gland and propeller. The propulsion system is made up of numerous components and each component has several variations and arrangements, depending upon the installation. It is important to identify which components are installed in your vessel and then to select the relevant sections from the information provided in this chapter.

BASIC PROPULSION SYSTEM

In simple terms, the propulsion system transmits power from the engine to the propeller. The number of components required and how efficiently it operates depends upon where the engine is located and the design of the system. There are a number of layouts for each propulsion system and these usually depend upon the type of vessel or engine installation. However, the combination of components will be fairly consistent, as the integral elements required by the majority of propulsion systems are the same.

It is worth reiterating that the propulsion system should be included as part of a regular maintenance routine – checking the condition of oils and operating cables before any journey commences can significantly reduce breakdowns and problems.

The basic propulsion system usually consists of the following elements:

- Drive plate
- Gearbox
- Coupling
- Prop-shaft
- Stern gear and propeller.

STANDARD PROPULSION SYSTEMS

The location of the engine in the boat has a direct impact on the type of propulsion system used. Applications where the engine is at the extreme aft of the vessel have the benefit of being less intrusive, thus providing more accommodation space. The further back the engine is situated, the steeper the shaft angle will be. Narrow boats are the exception to this rule. The result of a steep shaft angle is a loss of efficiency. More power is required to move the vessel, as there is an element of upward force as well as forward propulsion created by the steep angles. There are a few solutions that can be implemented to address this like Aquadrive and Python-Drive systems.

Conventional Propulsion Systems

In the conventional system, everything other than the propeller, prop-shaft and outer bearing is internal to the vessel and this means that it is accessible and can be maintained without the need to have the boat lifted out. This arrangement can be installed in most applications and is the standard arrangement on narrow boats.

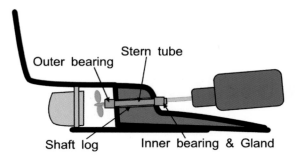

Typical conventional propulsion system, showing gearbox, prop shaft, stern tube and propeller.

V-Drive Propulsion Systems

These systems were designed to deal with installations where there is a steep angle caused by mounting the engine in the aft, or there are height or other restrictions that prohibit connection of an engine directly to a conventional propulsion system. The V-drive system can be directly coupled to the engine or mounted remotely. The key objective is to allow the transmission system to be virtually horizontal, therefore easing lubrication and component stresses. This will increase system lifespan and efficiency.

This type of installation can present some challenges when trying to access engine components mounted at the front of the engine if it has been placed close to bulkheads. In addition, access to the stern gland may be difficult and can be prone to oil contamination from leaking components.

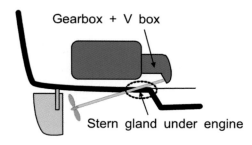

V-drive propulsion system – an alternative propulsion system, mainly used on vessels where space is at a premium.

Sail-Drive Systems

Sail-drive systems are popular in small yachts and cruisers. They enable an inboard engine to be used instead of an outboard. By utilizing a large rubber diaphragm positioned on the bottom of the hull, the drive leg and bevel sticks through and is submerged continuously in the water. This allows the drive to be mounted vertically, with the input and output shafts sitting horizontally in line with the hull. This offers many benefits to efficiency and drive, and is also compact and space-saving. This system is very rarely found on the canal system, but can be used on rivers; however, it is usual to find this type of system on vessels that are for coastal use.

The rubber diaphragm prevents water from entering the boat and does not require maintenance like a stern tube. However, the vessel has to be removed from the water to undertake any maintenance, or to investigate issues if it begins leaking. Having the correct anodes on your vessel will ensure that the section of the sail-drive system that is submerged is protected from corrosion and early failure.

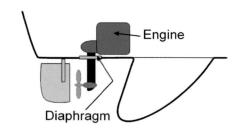

Sail-drive system – the typical system used on sailing boats.

Z-Drive Systems (Outdrive and Stern Drive)

The Z-drive systems are sometimes referred to by the manufacturer's name, so Enfield Outdrive, Volvo Outdrive, Mercuiser, Alpha and so on. As with the sail drive, much of the drive system is under the water, except that the drive is attached horizontally to the engine and exits the vessel through the transom (the rear of the vessel). This type of drive system contains many linkages and small parts, so when things go wrong it is always best to bring in a specialist. Even basic maintenance (just to change the oil in some cases) will require the outdrive or boat to be removed from the water.

All outdrive systems incorporate manual mechanisms for tilting the leg to remove the propeller from the water. This is useful for checking fouled or damaged propellers. Unfortunately, on older models these usually become non-operational due to wear and age, but a rope and pulley can be used in an emergency. The vessel steering is also controlled through the outdrive using a series of steering cables and these are prone to stretching and wear. If cables fail or become slack, the ability to steer will be affected, so regular checks and maintenance are essential.

The outdrive is also vulnerable to damage from hitting underwater obstacles, a problem that is especially prevalent on the inland waterways or shallow shores. If the outdrive hits something underwater, it can result in transom damage as well as internal damage and will require the outdrive to be removed, or the boat to be lifted out of the water for the damage to be assessed and repaired.

It is also worth noting that many outboards, Z drives and sail drives carry warnings about use of anti-fouling, as some can electrically attack the casing. It is recommended that they are painted and only anti-fouled with guidance.

Hydraulic Drive Systems

This propulsion system uses a combination of hydraulic pump, pipes and motor to transfer the engine power to the propeller. The benefit is that this type of system can be used regardless of the engine location, there are no vibration issues or torsional shocks and the gearing can be modified to suit all requirements by incorporating valves.

Age and wear are the biggest enemy of these systems, as wear will result in loss of efficiency and is typically seen as a reduction in performance and delays in gear change. Due to this, maintaining the oil and filters is vitally important. Failures and breakdowns will normally require a specialist engineer to attend, as hydraulic pipes are high-pressure components. These systems are becoming popular on the inland waterways.

DRIVE PLATES/TORSION COUPLINGS/CLUTCH PLATES

Between the engine and gearbox or outdrive there is usually a plate that is in place to absorb the torsional torque produced when the gears are selected and to dampen shocks from hitting underwater obstacles. These plates are referred to as drive plates, torsion plates or damper plates, but irrelevant of the name they all perform the same job. There are many different manufacturers and suppliers of these parts – R&D, Centa and Sachs (aka Sprigg plates) are some of the more common names.

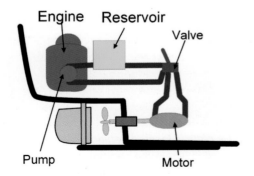

Typical hydraulic propulsion system.

A number of Sachs drive plates, which contain springs around the body.

A centa type of plate for heavy duty use.

A range of R&D plates, which are typified by the rubber bushes showing around the edges.

There are no maintenance requirements. However, due to the amount of stress and strain on these parts they do fail regularly and typically result in a loss of drive. They can fail catastrophically or over a period of time and will typically result in rattling, squealing or vibration before imminent failure. The most common symptom of drive-plate failure is a sudden loss of forward drive.

To access the drive plate for replacement the gearbox must be removed in most cases, although on some engines like Vetus, or where an adapter plate has been fitted, it is possible to access the plate by removing the bell housing with the gearbox still attached to it. Unfortunately, unless you have a record of the drive plate's part numbers, it is usually necessary to remove the gearbox and drive plate in

order to identify the type and model. The drive plate is bolted directly to the flywheel with a ring of bolts fixing them in place. There are so many drive-plate types in circulation that it is rare that a specific gearbox and engine configuration will have the same drive-plate arrangement.

Removing a gearbox will depend upon the engine type and arrangement and it is best to refer to specific manuals for guidance on how to do this. Alternatively, have an engineer attend to undertake the work on your behalf. This section will provide guidance on what to be aware of should you opt to remove the gearbox to inspect and replace a drive plate.

The drive plate is located behind the engine, inside a bell housing where the gearbox is connected. To access, the gearbox will need to be removed and in some cases the bell housing.

A failed R&D drive plate. Note the rubber bushes have sheared and are no longer whole.

TIP: DRIVE-PLATE ISSUES

When a drive plate starts to fail it can break up and sections of the plastic damper can escape the bell housing and get lodged in the starter motor. This is a common issue on the Vetus engine and can be the first indication that the drive plate is failing.

TIP: GEARBOX GUIDANCE

When changing a gearbox, always ensure that it matches the one removed. If upgrading or modifying, it is essential that the propeller and engine size are checked to ensure that the correct power output is maintained.

GEARBOXES

The purpose of the gearbox is to convert the engine power safely into horizontal movement and thus provide astern, ahead and neutral through gear selection. The gearbox is normally coupled directly to the engine, unless a hydraulic system is used. The gearbox ratio will dictate the amount of power transferred to the prop-shaft and directly affect propulsion through the water. The gearbox ratio is matched to the size of the propeller, vessel size and power produced by the engine and in many cases the use of the vessel.

Although we do not cover gearbox internals in depth in this book it is worth mentioning that marine gearbox internals are significantly different to vehicle gearboxes. There is no external clutch to disconnect the drive from the input shaft of the gearbox, so the gears are always engaged and therefore continuously moving when the engine is running. They do contain an internal clutch, but due to the space restrictions this is small and can only engage gear safely when the engine rpm are low. Certain types of gearbox need to have neutral selected prior to engaging gear in order to reduce potential damage to the gearbox. Some manufacturers may have warnings regarding excessive running out of gear, high rpm gear engagement, or going straight from one gear to another without a pause. Some gear controllers (Morse controllers) may be fitted with an electrical switch to prevent the engine starting unless the gears are in the neutral position for safety purposes.

Ratios are normally in the range 1:1 up to 4:1. This is because a vessel with the same hull and engine combination can be used for different applications. A vessel that is used for offshore angling may require a 1:1 reduction to allow it to motor at high speed out to a fishing ground. However, an inland vessel with the same set-up might require a 4:1 reduction to provide maximum power at low speed, although the majority of boats are fitted with 2:1 ratio gearboxes.

Reduction Gearbox

Some installations, in particular vintage engines, may have an additional reduction gearbox, fitted in line with and as part of the main gearbox. Such

The location of the identification plate and details on a Hurth gearbox. This is a HBW 100 2R (2R = 2:1 ratio).

A PRM mechanical gearbox showing the location of the identification plate, with the information on it indicating PRM80D2, and the filler and dipstick (D2 = 2:1 ratio).

an installation will require the oil to be topped up separately from the main gearbox and, in some cases, will use a different oil grade. However, apart from the oil top-up and change there is no maintenance required. Both gearboxes will have a separate dipstick located on the side of the boxes.

There are many different types of gearbox and it is important to identify the type that is installed in your vessel. This information can usually be found on a printed plate on the gearbox and will contain a serial number, model number and manufacturer, along with the ratio. In this section, we look at the different types of gearbox, typical manufacturers and the components that interact with the gearbox.

Mechanical Gearboxes

Common manufacturers of marine gearboxes include PRM, Hurth, Twin Disc and ZF. There are also many older gearboxes, for example Borg-Warner and HyDrive, and many of the bigger engine manufacturers produce their own gearboxes. The older gearboxes can be difficult to obtain parts for and, in many cases, will need to be rebuilt by a specialist and may need to have parts custom-made to complete repairs. If a replacement gearbox is required, it is essential that the gearbox ratios,

the propeller size and the engine horsepower are matched to ensure that the power ratios are suitable for the current arrangements. The operation of older gearboxes or manufacturer-specific ones are not covered in this book.

The mechanical box refers to the clutch mechanism within the gearbox. Manual gearboxes historically utilize two types of clutch systems: a clutch pack consisting of a series of friction plates engaged with a large spring; and a cone-type configuration that meshes directly with the gear selected. An external lever is used to select either forward or reverse gear. Mechanical boxes may be quite abrupt when engaging gear, which places a great strain on internal components and can result in damage occurring internally. It may also result in shorter drive-plate life.

The biggest cause of failure is due to clutch slip, which occurs with age, wear and misuse. There is usually a warning that this is occurring as the gearbox will start to slip or jump in and out of gear, or may begin to engage gear increasingly abruptly (a loud bang when selecting gear).

Mechanical boxes require the cable to be adjusted at both the Morse controller end and the gearbox end. This should be part of a regular maintenance

A Hurth gearbox is a mechanical gearbox. Seen here are the bolt for the filler and dipstick.

A Twin Disc mechanical gearbox, showing the location of the dipstick and the filler bolt. Note the adjustment positions on the arm.

routine, as when a cable is poorly adjusted it can act like a slipping clutch in a vehicle and directly affects the life and wear of the internal components of the gearbox. This can be checked by ensuring that the selector arm sits centrally when the controller is set to its neutral position.

Regular oil changes are essential to maintaining the life of a mechanical gearbox and the appearance of debris or metal filings in the oil is not unusual if servicing has been sporadic or left too long. A new gearbox should have the oil changed within 25hr and the selector arm adjusted – as the gearbox 'beds in', the arm will need to travel further to engage the gears correctly and will need adjusting periodically. The oil service period should be outlined in your owner's manual, but as a general rule the oil should be changed as soon as it begins to become discoloured.

Hydraulic Gearboxes

The hydraulic gearbox uses a hydraulic clutch system, which provides a smoother gear change and puts the internal components under less stress. There will be two clutches, or clutch packs, internal to the gearbox, which utilize friction plates that are engaged using a hydraulic piston. The gear change is generated when oil is directed under pressure to either one of these clutch packs, although externally the mechanical cable and arm system are still used. These gearboxes are generally more reliable and in many installations can be over thirty years old without ever experiencing any issues, as long as they have been regularly maintained.

Showing the gearbox selector arm and adjustment positions that the cable can be fixed to in order to achieve the correct movement.

A PRM hydraulic gearbox showing the location of the identification plate, which indicates that this is a PRM 150D2.

The location of the drain bolt, although this may not be accessible (it is usually a brass bolt). If not, you may need to use an oil extractor through the filler hole.

PRM 160 hydraulic gearbox, showing the identification plate and the filler. Note that the yellow caps are where the hydraulic pipes are located.

The disadvantage of the hydraulic gearbox is that if the oil pressure is lost, drive and propulsion will be affected and in severe cases the gearbox will not engage gear. Due to this, it is important to check regularly for oil leaks, as most hydraulic gearboxes have issues with oil seals failing.

Showing where to connect the hydraulic pipes to the gearbox.

Slip still occurs in these gearboxes, but it is usually due to contamination and oil degradation, resulting in hydraulic pump wear, which affects oil pressure as the clutches use oil pressure to create the friction required. Any loss of pressure will result in clutch wear and eventual failure. Some hydraulic gearboxes also have mechanical adjustment, which, if incorrect, will result in gearbox slip. In some types of gearbox, like the Lister, this issue has been addressed by incorporating an emergency spring, which, when pressure is lost due to low oil or band failure, locks the gearbox in the ahead position so that the boat can still move.

Gearbox Oil Cooler

The majority of mechanical gearboxes will not have any form of oil cooling, but all modern hydraulic gearboxes require oil coolers to be fitted. This is because the oil absorbs the heat from the internal components like the friction plates and if not cooled would degrade extremely quickly and fail to transfer heat efficiently from the major internal components, quickly resulting in gearbox failure. Some very old gearboxes will also incorporate an oil filter as part of the gearbox. The oil cooler is constructed with an inlet and outlet for both the water and gearbox oil. The cool water passes through the internal copper tubes and so the gearbox oil that surrounds the tubes is cooled.

The cooling system provides cooling for the engine and the oil cooler, and is usually the first component in the cooling system. Unfortunately, for

Construction of a gearbox oil cooler – the internal tubes have water running through them to cool the surrounding oil.

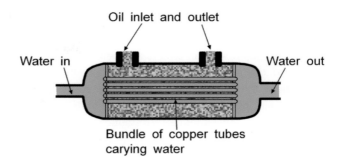

coolers installed in raw-water-cooled systems, the sea-inlet strainer, which prevents debris entering the engine cooling system, can let a certain amount of fine debris pass through the water pipes of the oil cooler. This rubbish eventually blocks the water tubes and can result in the cooling system failing. It is therefore advisable to remove the water hoses regularly and remove any rubbish from the water-inlet side of the oil cooler. Reverse flushing with a hose is the quickest – if messiest – method. On newer engines, the oil cooler is sometimes incorporated within the heat exchanger for the engine.

Gearbox Service

A regular routine of checking the selector lever adjustment, changing the oil and topping up the gearbox should result in years of trouble-free operation. The oil should be checked every few months,

Small oil cooler, usually used for cooling systems on gearboxes.

or before every long journey if accessible, and topped up as required; developing a regular routine will highlight changes in oil condition or usage.

You will need to identify the dipstick location on your gearbox. There may be more than one, as some types have a number of reservoirs (Lister, for example). Locate the oil draining point, oil type and filter, if there is one. Note that older gearboxes use oil filters more so than newer types. Following is a guide for a standard gearbox service:

- If the gearbox has an oil drain, loosen this and leave the oil to drain into a pan placed under the gearbox. (Place an oil-absorbent sheet underneath to reduce the likelihood of oil getting into the bilges.) The oil drain will be located at the bottom-most point on the gearbox casing and

Locate the oil drain plug on the gearbox. It is usually brass and at the bottom of the gearbox, but is not always accessible.

is usually plugged using a brass nut. Consult the manual for further information on its location. If the gearbox drain plug is inaccessible or damaged, use an oil-siphon pump to remove the gearbox oil through the filler hole; just be aware that this will not remove the debris in the bottom and is therefore only a temporary solution.

- Once the oil has been drained off or extracted, if the gearbox has an oil filter it will be located centrally on the top of the gearbox casing. Twist this off, catching any remaining oil (have a towel handy) and place in a tray. Remove the filter seal and clean inside the rim (if it is a gauze filter, it may be inside the dipstick hole).
- Fit a new seal and replace the filter with a new one (follow the same process as changing the oil filter when servicing the engine). Refit the drain plug, which was removed to empty the gearbox oil. If you are unsure of the type of oil to use, refer to the manufacturer's handbook, or the nameplate. However, most hydraulic gearboxes take engine oil and most manual gearboxes take ATF oil.
- Fill the gearbox using the manufacturer's guidance. However, if you are unsure of the correct amount, the general guidance is to fill the gearbox to the lowest marker on the dipstick if it is a screw-in type, or just below the highest marker if the dipstick is a push-in type. If during regular inspection the oil is found to be either creamy or to have a thin consistency, there may be a chance that the oil cooler has failed, or a seal in the gearbox or oil cooler has failed. If this happens, the oil will need to be flushed and replaced, and the cooler inspected and or replaced.

Gearbox Removal

It is not advisable to attempt to remove a gearbox unless you are extremely confident. However, if you decide to attempt it, for example to inspect the drive plate or to change the gearbox, you must consider the following:

- The prop-shaft will need to be disconnected from the gearbox and pushed back to provide room for the gearbox to be removed.

TIP: GEARBOX MAINTENANCE

If a gearbox is ever submerged, due to water in the bilges or water ingress into the boat, it is always worth flushing and changing the oil, as water will migrate through rubber seals and contaminate the gearbox oil. The effect of water in the oil is that the internals of the gearbox will quickly deteriorate and fail.

- The coupling may need to be removed from the prop-shaft if space is at a premium. This is especially true if a Centra coupling is fitted.
- The cable linkages will need to be removed from the gearbox and placed away from the working area. These will need to be readjusted when refitted.

Gearbox attached to the bell housing on a Vetus engine.

- Any brackets will need removing and repositioned (these may be electrical cable harnesses or relay mounting brackets) to allow space for the removal of the gearbox and possibly the bell housing.
- The starter motor may need to be removed, as this will be bolted into the bell housing.
- The engine may need to be supported if the engine mount brackets need to be removed. You can use a scissor jack or sling to accomplish this. If you fail to do this correctly, the engine will drop into the sump, so ensure that the support is safe.
- The engine mounts may need removing in order to shift the engine back; this will result in the engine needing to be realigned when the gearbox has been reinstalled.

COUPLINGS

Anti-torsional vibration couplings are characterized by the use of rubber inserts. They are quite stiff; in fact, they are usually impossible to bend by hand. They smooth out the torque from the engine and cushion it from any underwater impacts that occur. This type of coupling must always be aligned to the engine and prop-shaft or it will result in vibration, which in turn can shear bolts or damage the coupling, causing issues with the prop-shaft and engine mounts.

R&D Couplings

These are one-piece flexible couplings that allow a limited amount of angular and radial (vertical/horizontal) misalignment. To remove these couplings, remove the eight main bolts that connect them to both the gearbox and the prop-shaft. These couplings are relatively inexpensive and are designed to be sacrificial should the vessel impact an underwater object. The mounting points have a habit of wearing and the metal sleeves fall out when wear is excessive. Check for this by removing once a year and replace the coupling if the sleeves are loose.

Industrial Flexi Couplings

Centaflex and Fenner typify this type of coupling. They are designed so that if a section fails, the collar can be replaced without needing to replace the whole coupling. Never exceed the amount of misalignment quoted by the manufacturer. The gearbox and shaft have to be aligned with this type of coupling and it is important that they are checked every year. (A solid dummy coupling is ideally required when checking alignment of the shaft.) To remove this coupling you will need to remove the eight bolts on the prop-shaft side. Next, reinsert into the four alternative bolt holes, which will disconnect the coupling from the shaft (sometimes there are only two holes). Follow this with the removal of the four nuts on the gearbox side. This coupling type is moderately expensive, although replacement parts are less so.

R&D coupling – 4in.

Centaflex coupling – side view.

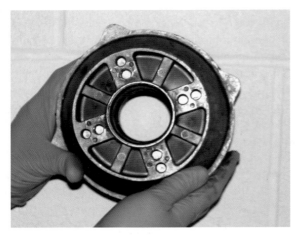

Centaflex coupling – end view.

Fenner-type coupling with rubber boot.

TIP: PREVENTING COUPLING ISSUES

If the coupling bolts come undone on any coupling, always replace the nyloc nuts when refitting the coupling, as they will continue to vibrate loose if not and will eventually result in the complete coupling needing to be replaced.

Aquadrive and Python-Drive Systems

These are flexible jack-shaft types and are typified by the Aquadrive and Python-Drive systems, although there are many different arrangements that use intermediate shafts containing a pair of vehicle universal (CV) joints. These are the only 'fit and forget' couplings as far as alignment is concerned, as they can cope with up to 16-degree misalignment. However, this can lead to issues, as prop-shaft alignment has to be within the specified angle for these products to work and exceeding the alignment angle will result in early failure.

These couplings are very expensive, but if installed correctly are the most robust and reliable on the market. They might need greasing regularly, esepcially if they are universal joint (UJ) types,

which will have a grease nipple fitted. This system provides better support for the propeller shaft, resulting in smoother running and less wear on the stern gear. It is common to see these still in operation after more than ten years.

If the alignment is extreme or the prop-shaft is bent, or following an impact, the seals may begin to leak. The seals on these coupling can be replaced in situ, although as the prop-shaft has to be moved back to get access it is usually recommended that a qualified mechanic undertakes this work.

These systems require a thrust block to transfer prop thrust to the hull, otherwise they will 'dog-leg'. The thrust block might be part of the assembly (Aquadrive/Python-Drive), otherwise the thrust block will also need regular lubrication.

CV joint, the basic concept behind both Python-Drive and Aquadrive.

Aquadrive coupling installed in a narrow boat.

Coupling Installation Considerations

When considering how the coupling should be fitted to the shaft you need to consider:

- The cost of the shaft.
- What happens if the shaft gets bent.
- How likely it is that the shaft will get bent.
- How easy the coupling is to remove so that the shaft can be withdrawn without removing the engine.
- The length of the prop-shaft and if any additional support couplings are used.

The ways that couplings are fitted to shafts fall into the following categories.

Industrial: These were originally designed for use in machines and are characterized by a number of socket screws or large collet nuts around their circumference to tighten the coupling to the shaft. These are easy to fit and to remove (providing they have not gone rusty) and are very popular.

Shrunk-on: These have a slightly smaller hole in the coupling than the diameter of the shaft. The coupling

is heated to expand it, so that it (in theory) slides over the shaft. These are often bored out to a sliding fit and a single bolt is drilled into the shaft to take the entire prop thrust. A properly shrunk-on coupling is fixed for life, so is very difficult to remove. These do not respond well to bent shafts – they work loose.

Taper pin: The coupling is bored, vertically, to one side of the shaft and the hole reamed to a taper. The hole is half in the shaft and half in the coupling. A taper pin is then driven into the hole. The pin tends to push the coupling to one side, especially if the bore or shaft is worn, so that it is impossible to align radially.

Taper fit: Both ends of the prop-shaft are machined to standard Admiralty taper (as if the prop was going to be fitted to both ends) and the half coupling machined so that it is also tapered and has a rebate to accept a nyloc nut. By slackening the centre bolt and trying to 'refit' the half couplings the gearbox flange nut will usually break the taper so that the coupling is easy to remove. The bearings and stern gland will wear the shaft over time, but usually by reversing the shaft and putting the unworn surfaces into the bearings and gland, the shaft life can be doubled.

Ideally, all couplings should also be fitted using a key to transmit the torque. Woodruff (semicircular) keys are more reliable than square keys.

STERN GEAR

There are many different types and arrangements for the stern tube, although unless you are unlucky enough to need to have the stern tube replaced the only maintenance needed is to grease and replace the packing regularly when it begins to wear. It is different for water-lubricated stern gear (*see later*).

The stern tube provides the access point for the prop shaft to be connected to the propeller on the outside of the hull. Therefore, it has to provide a lubricated but water-tight environment for the prop to turn without causing friction, heat or movement. The conventional arrangement on a narrow boat and most cruisers is to use stern gland packing, which is wound around the shaft and then compacted down with the gland follower. Water flows inside the tube and the packing acts as a seal, but a small amount of water will always get through and this is why it is perfectly normal for a stern tube to leak when the vessel is moving. However, if it leaks when at rest there may be an alignment issue, or the packing may need to be adjusted or replaced.

As the packing wears, the amount of water that escapes will increase and it is important to check

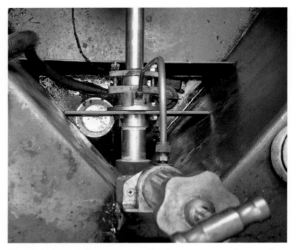

Stern tube and greaser – the brass pump is used to pump grease into the stern tube to lubricate it. At the end of each journey, pump grease into the stern tube until the water stops dripping.

frequently to identify if there is a change. The stern tube should be greased regularly to keep the system lubricated and extend the life of the packing. Before and after every trip, grease should be pumped into the stern tube using the grease connection on the top of the stern tube. One turn of the greaser should be sufficient for up to eight hours of cruising.

Maintenance: If the stern tube is leaking more water than normal, this indicates that the packing may be wearing and needs to be compacted further. To achieve this:

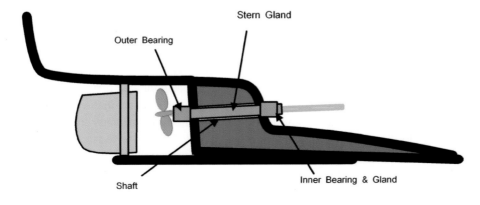

Main components of a typical stern tube arrangement.

- Locate the adjustment nut or flange on the engine side of the tube. There may be two nuts on opposing sides of the flange, along with two locking nuts used to stop free movement. Alternatively, there may be a single large nut encompassing the prop-shaft along with another locking nut to stop free movement. The locking nuts should be loosened to enable movement before adjustment and retightened once the adjustment has been accomplished.
- The adjustment nuts should then each be tightened equally a quarter of a turn at a time to compact the packing,
- If you cannot turn them, the packing may be at the end of its life and may need to be changed.

Stern gland nut single nut. Undo the main screw, screw in the end section to compact the packing, then tighten the nut to lock into position.

Stern gland nuts with a double-nut arrangement. What is done on one side should be done on the other, so that the packing is compacted evenly. When there is no adjustment nut left, this means that the packing needs changing.

The prop-shaft should always spin freely in the stern tube and you should feel only the tiniest resistance when spinning it by hand. If there is lots of resistance and still an excessive water leak, the engine is likely to be misaligned. Once the adjustment has been completed, check that the prop-shaft is free and turning without any tight spots.

TIP: ADVICE ON STERN TUBE COOLING

How much should a stern tube leak? There is not a standard answer to this and, more importantly, the question should be: Is the right amount of water flowing around the tube? The best way to check this is to feel the stern tube after a cruise. If it is too hot to touch, the amount of water flowing is too little and the flange needs to be loosened to allow more water to flow. If it is cold to the touch, then it is probably too loose and the flange needs to be tightened.

Sealed Stern Tube (Mechanical Seals)

The other type of stern gear that may be installed on your boat is typified by the Vetus sealed stern gear, which is a completely different arrangement and importantly if you have this type of installation on your boat, it should not leak any water. These types of systems benefit from the fact that there is no leak and therefore bilges remain dry.

Sealed stern tube (mechanical seals).

All the systems produced by manufacturers like Volvo, PSS ManeCraft and Vetus split into two main categories – face seals and lip seals. Both systems use a long rubber tube to keep the water out, but use different methods to seal on to the shaft.

Replacing Seals on Sealed Systems

In general, these sealed systems are extremely refined and robust if a few maintenance tips are observed. Although these are water-lubricated, the seals should be treated with respect. Misalignment of the engine will cause the seals to leak and damage the self-aligning bearings inside the gland, so it is always advisable to check the engine mounting nuts for tightness.

Good practice is to maintain the seal flexibility by regularly applying silicone grease to the seals, about

every 300 to 400 hours. There are various methods to grease the seals depending upon the manufacturer; referring to your manual will give guidance on the best method to employ. The most common ones are detailed below, but the key here is to use the correct type of grease – silicone grease should always be used, not standard grease:

- Vetus – remove a hex grub screw located above the seals and in front of the water bypass pipe, apply grease, then replace the screw.
- Volvo – squeeze the seals and insert a tapered tube into the misshaped seals, apply the grease, then remove the tube.
- Radice – Pop out the rubber bung. Remember to replace parts that were removed after greasing.
- All three of them need a squirt of grease big enough to cover a two-pence piece.

Following is the procedure to replace seals:

- Wrap a greasy cloth around the shaft and stern tube on the outside of the boat – through the weed hatch – in order to stop the majority of water ingress while removing the seals.
- Remove the propeller shaft coupling and push the shaft to the stern (not past the seals – preferably 5cm of shaft should be visible).
- Remove the seals. This can be accomplished by either removing the seal plate or removing the rubber hose clips and removing the front seal:
  - In the case of Volvo or Radice, replace the complete inner gland; remember to tighten the hose clips.
  - In the case of a Vetus gland, remove the seal set with the three hex screws. Check the bronze bearing for wear, for example ovalling, and replace all worn parts. Be sure to use the seal-saver and fill the cavity between the two seals with silicone grease when replacing the inner gland.
  - Unfortunately, Vetus does not provide a seal saver with the replacement seal set, so care is needed when installing.

- Refit the shaft into the coupling. Check the engine alignment at this point. If the propeller shaft has grooves where the seals have run, move the shaft into a different position so that the seals sit only upon unworn shaft. Sometimes it may be wise to change the shaft. However, if replacing the Vetus gland with either Volvo or Radice, the complete gland is shorter and the seals then would run on unworn shaft.

PROP-SHAFT

Prop-shafts are basically a long metal bar with a tapered end, which allows the components that make up the drive system to be attached to a common location. It transmits the drive and therefore must be able to deal with the torsional forces at work during both normal operation and should the vessel hit an underwater obstacle.

The prop-shaft only suffers with a couple of issues: one is the result of impact; the other poor engine alignment, which results in the prop-shaft becoming bent or excessively worn. An impact usually originates from under the water. However, when fitting engines there is also the risk that this can occur if care is not taken. Misalignment will cause the prop to become bent over time. This can occur when an engine leg or engine mount has failed and the engine has sat for long periods. This can also be the case if incorrect engine mount adjustment is present. In both situations, the bend will result in ovalling of the stern tube, vibration and harmonic noise.

The other issue is usually caused by components like couplings and taper locks that are attached to the shaft and which work their way loose or spin on the shaft. This does not affect the usability of the prop-shaft, but can cause wear that affects the ability to fit parts to it and if there is nowhere else to fit the component it can result in the shaft having to be replaced.

Prop shaft showing the tapered end where the propeller sits.

TIP: IDENTIFYING ALIGNMENT ISSUE

To check the prop-shaft condition, turn it by hand. There should be no stiff points, grating or tightness. The prop-shaft should turn smoothly and in one rotation; if it does not, call in a specialist to check the installation.

PROPELLERS

Propellers come in many sizes and shapes and new developments are continuing to evolve and achieve better efficiency and drive. It is rare that you will need to change a propeller, unless you damage or lose it, or are changing or upgrading the engine or gearbox. However, if your vessel is out of the water, it is worth investing the time in taking some details down about the size and checking its condition. Propellers on the inland waterways suffer more frequent incidents and impacts and, therefore, damage to the edges and bending or distortion of the blades is common. These impacts reduce efficiency, affect drive and can result in vibration and harmonic noise travelling through the drive system. This is commonly known as cavitation.

A propeller is defined by its diameter (in inches) and the pitch (in inches), and will typically be something like 18×12 LH or 18×12 RH, where LH and RH indicate the rotation, with clockwise being right-handed and anticlockwise left-handed. The first number is usually the diameter and the second number is the pitch.

The pitch represents the angle of the blade to the shaft. The measurement of pitch is derived from the amount the blade travels on a single rotation. The easiest way to see this in operation is to spin a propeller on its shaft and rotate it by one revolution; the distance the blade travels is the pitch measurement.

Typical propeller damage due to hitting an underwater obstacle, which can cause excessive vibration and result in noise and component damage.

CHAPTER 6

OTHER COMPONENTS

MORSE CONTROLLER

The Morse controller on your vessel will normally perform the dual operation of controlling the throttle (revs) and gear selection. To engage the gears and move the vessel, you move the lever from its neutral position to ahead or astern. This should be done in a single fluid and smooth action to enable the gears to engage positively; gently easing into gear can actually cause issues, a bit like 'dragging' the clutch can cause a burnt-out clutch in a vehicle.

Depending upon the gearbox, you may need to move the lever into the neutral position prior to changing from ahead to astern or vice versa. This is typical on manual gearboxes like the Hurth, as the harsh nature of gear changes can cause damage to the gears and eventually lead to slip or failure. It is generally good practice in all cases and will result in a longer life, even if you simply pause in neutral for a second or so before selecting gear.

The throttle control on the Morse controller can be in many forms, depending upon the type of Morse controller installed. On most canal vessels, the Morse controller will be a single lever, whilst on cruisers and some wide beams the Morse controller may be a twin-lever arrangement. In addition, if there is a second helm position or second engine, there may be an additional set of controls to consider.

Single-Lever Controllers

With the single-control lever, the movement of the lever arm performs two actions: the first is to engage the gears; and the second is to apply revs to the engine. The Morse controller has two cables attached internally, one controlling the gear selection and the other the throttle. The initial movement from the neutral position engages the gears. As the lever travels further, the throttle is engaged from an idle position and the revs rise, dependent upon how far the lever is pushed. The same rule applies regardless of whether ahead or astern is selected.

There are many different designs of single-lever Morse controllers and all will have a method to allow the engine throttle to be operated without having to engage the gears. Some will have a button that is pressed and held in to disengage the gears and ensure that they can't be accidently engaged. Others are pulled out away from the main body to unlock and allow independent operation of the throttle.

The biggest issue affecting the operation of the Morse controller is usually due to poor adjustment of the cables and the connections to the Morse controller. This can occur over time as the cable stretches, or through vibration causing the fixing and holding bolts to loosen. If the cable adjustment is incorrect, it will cause the gears to engage incorrectly and can lead to gearbox wear, which can result in the gear selection feeling 'lumpy'.

Checking, inspecting and adjusting the Morse controller will be dependent upon the type and how it has been installed and where it is positioned. If you are lucky enough to have a simple installation with good access, follow the procedure below. However, if not you may have to spend some time dismantling and accessing the internals, but once you have the Morse controller open the components and methods detailed below should still be applicable.

Twin-Lever Controllers

These controllers are a lot simpler than the single controller. However, due to space restrictions they are rarely installed on narrow boats. The twin levers operate the throttle and gears separately. With each lever operating a push-pull cable independently, it becomes much easier to engage gears without mistakenly adjusting the engine speed. These are, however, much more robust and require far less maintenance than the single-lever varieties.

CABLES

Cables are used to transfer movement from one component to another, usually on a push-pull basis. They are normally controlled by a lever or button at the user end and connected to a lever at the other end, which then operates machinery.

On a boat, the gear cable is operated at one end by the Morse controller, which transfers this movement to the gearbox, where it causes a corresponding lever to move selecting gear. The throttle is connected from the Morse controller via the cable to the injection-pump throttle linkage, where movement controls the amount of fuel delivered to the pump, thereby increasing or decreasing engine speed.

The key to cable operation is maintenance. Regular greasing of the ends, changing when the cable needs replacement and ensuring that the cable is routed correctly will provide trouble-free cruising. A cable should never have a bend in it of more than 90 degrees; if it does, this will create a 'tight' spot that will affect the operation of the cable and result in early failure. A cable, where possible, should fit the installation – if it is too long, it will be looped, but this will lead to more wear and tear due to additional internal friction.

It is always advisable to carry a spare cable. In most cases, other than for outboards or outdrives, the cable can be used for both throttle and gear and therefore only one spare is required. Next time you change your cable, record on the Engine Details sheet (*see* Appendix) the length along with other details if it is a specialist cable.

TIP: CABLE ROUTING

If the cable route is difficult or requires dismantling of parts of the boat, route the spare along the other cables, then if one fails you can quickly and easily connect the ends and be on your way. Alternatively, tape the old cable end to the new one and pull through.

Adjusting, Checking and Replacing Cables

If you suspect a cable failure, first check for movement at the gearbox or throttle end, as if there is

Morse controller with the front cover removed.

movement it may indicate a different issue. To access the cable, remove the front panel of the Morse controller by removing the fixings – taking care not to damage them. If the screws look worn or damaged, replace them with new, as these items usually sit open to the elements and corrode over time. If reusing the old ones, smear a small amount of grease or petroleum jelly on to them before refitting to stop corrosion and to make removal in the future easier.

Inside the body, you will see two cables usually routed through the bottom of the controller, although sometimes they come from the back. Trace the cables and you will find that they are both attached to separate elements within the body. One controls the gears, the other the throttle. Make a note of which cable goes to which pivot point (tag them or take a photo if you have time). Check the operation by moving the Morse controller forward and back, with throttle engaged and not engaged. Ensure that the cables and connections are giving full movement.

If the Morse controller is functioning correctly, but is not resulting in movement at the gearbox or injection-pump throttle assembly, this means that the cable has failed and needs replacing. However, if movement is restricted, or the cable is not being operated correctly, investigate further. Typically, the

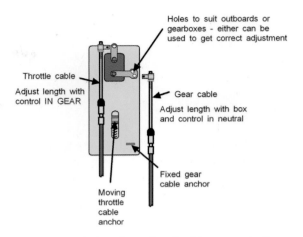

How the cable is adjusted at the Morse controller end.

pivot and saddle clamps that hold the cable work loose, so check that they are fitted correctly and seated in the correct location. Once you have identified and resolved the issue, always test that the installation is working correctly before refitting the main plate.

Changing a Cable

If after checking the cable you identify that it needs to be changed, you will need to locate any cable clamps and release the cable prior to disconnecting it.

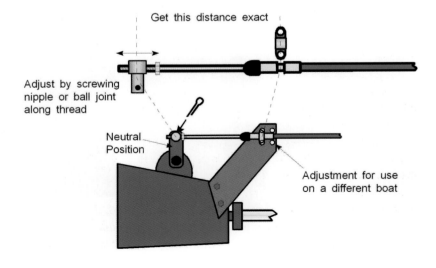

How the cable is adjusted at the gearbox end.

1. Locate the cable clamp bracket.

2. Lift the clamp to release the cable (do not forget to refit once the new cable is fitted).

3. To disconnect the cable from the gearbox selector arm, remove the split pin that holds the cable in place. You may find there is a spacer, nut or clip that also needs to be removed.

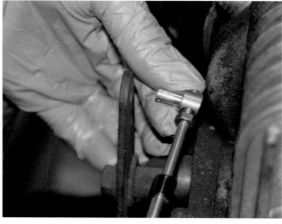

4. Slide the cable end out of the hole on the gearbox selector arm.

5. The cable adjuster sits on a threaded section with a holding nut below it to prevent movement.

6. To adjust the cable, loosen off the holding nut, turn the adjuster to the desired location, then tighten the holding nut to secure.

7. *Insert the cable back into the gearbox selector arm, then test the operation of the Morse controller and the movement of the selector arm. If required, perform again until a good range is reached.*

To disconnect the cable, remove the pin/nut/connector at the injection pump or gear lever on the gearbox (take a photo or make notes on its arrangement for reference). The following will guide you through the process to release two different cable ends at the gearbox:

- To access the cable at the Morse controller end, if not already removed you will need to remove the front plate of the Morse controller to gain access to where the cables are located.
- Locate the cable ends and remove the split pin holding the cable in place; you may need to move the Morse control lever to gain access if required (take another photo for reference). Do one cable at a time.
- Disconnect the cable from the Morse controller.

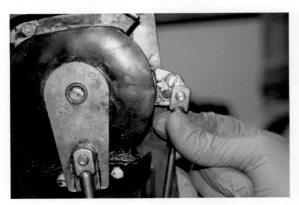

1. *Cable split pin removed and the cable is ready for removal.*

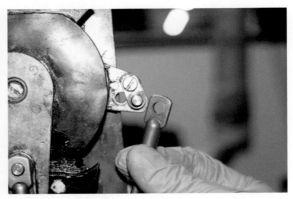

2. *Cable removed from the holding pin on the Morse controller.*

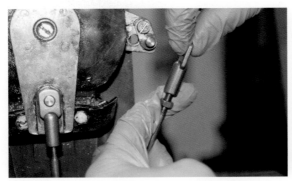

3. *Twist the cable end to adjust the cable reach, moving the end up and down the threaded end of the cable.*

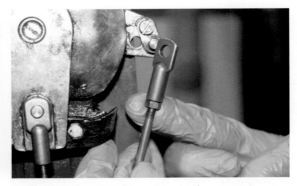

4. *Tighten the securing nut once the correct adjustment is reached, then refit to the Morse controller, leaving it in the neutral position.*

- If access is good, remove the cable and then route the new cable along the same route, ensuring that there are no bends greater than 90 degrees. Remove any removable components from the old cable and refit to the new replacement cable.
- Connect the cable at the Morse controller end, making adjustments as shown in the photograph, and test that the installation is working correctly before refitting the main plate. The cable ends are threaded to allow adjustment and winding in or out the cable connector allows for positional change.
- The best way to guarantee that the cable is adjusted correctly is to have the cable connected to the Morse controller and the controller in the neutral position. Then refit the cable to the engine side by reversing the operation to remove them (using the pictures or notes as a guide). On the engine side adjust the end so that it connects to the throttle arm when in its rest position. The throttle arm should automatically sit in the correct position with the cable disconnected. On the gearbox selector arm it should be positioned centrally.
- Make sure that the cable is adjusted as shown in the photographs before fitting.
- Check the operation of the gears and make sure that the gear selector arm on the gearbox is travelling the same distance in both directions.

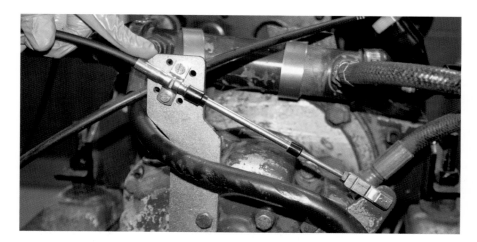

1. An alternative arrangement is shown in this photo, with a permanent clamp and sliding clip that secures the clevis cable clamp.

2. To remove the cable, slide the clip along the clevis jaw to reveal the end.

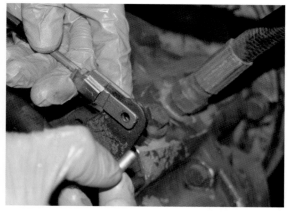

3. Remove the holding pin from the centre and release the cable.

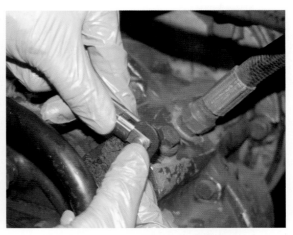

4. To adjust the cable, twist the clevis jaw on the threaded cable end until the desired position is achieved.

5. Insert the cable back into the gearbox selector arm, then test the operation of the Morse controller and movement of the selector arm. If required, perform again until a good range is reached.

- Start the engine and check the throttle, adjusting the throttle at the injection-pump end of the cable if required (if the rpm are higher or lower than when you began).
- Check the cable operation – this should be smooth and fluid. Any drag or slip is a sign that the cable is not routed or adjusted correctly, in which case adjust the cable and test again.

ENGINE MOUNTS

Virtually all engines will be fitted with 'flexible' engine mounts. These are referred to as flexible, as they incorporate a rubber shoe that is attached to a metal cup or plate. The main bolt that is used to adjust the height of the engine runs through the centre of the rubber.

The rubber reduces vibration and noise and helps to keep the engine stable and unaffected by movement. However, over time the rubber deteriorates and it can become compressed and rigid. Alternatively, if contaminated with oil or diesel it will become soft. In both cases, it will result in increasing vibration and engine noise. These symptoms will affect the engine operation as they put strain on other components and eventually cause failure.

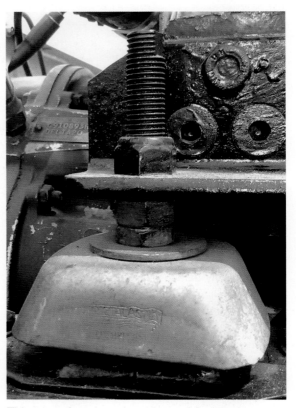

This type of engine mount is used for heavier engines. The rubber section is protected by the metal case.

This type of engine mount is used for lighter engines.

Engine mount showing deterioration of the rubber component, which indicates that the mount is ready to be changed.

- The bolt has sheared.
- The adjustment nuts are worn or continually come loose.

If you identify an issue it is good engineering practice to change all engine mounts rather than just one; changing the two rear or two front mounts will suffice, but you are taking a risk. Depending upon how long the engine mounts have been in operation there will be significant difference in the old mounts' ability to absorb vibration. This puts extra load on the new mounts, shortening the life expectancy.

Adjusting and Tightening Engine Mounts

The engine needs to be in the correct position to maintain its optimal operation, but getting the engine aligned can be complex and time-consuming and is usually best left to a professional. However, there are times when an issue with the engine mount needs to be addressed. The nuts can become loose due to the natural vibration from the engine and knowing how to tighten them is crucial.

On each engine mount there is a nut on the top of the bolt and one underneath connecting the engine leg. Tightening the engine mounts is a simple process of winding up the bottom nut on the engine mount until it becomes tight – however, *never* adjust the top nut, as this will affect the alignment of the engine and can cause catastrophic damage to other components. If the top nut has moved, it is not advisable to attempt to guess the position and a professional should be called.

If not replaced, the engine mounts can eventually fail, or engine mounting brackets (engine legs/feet) will shear due to the increased vibration or movement. The vibration can also cause fuel pipes to shear, impact upon alignment and cause wear on the stern tube and couplings.

Engine mounts should be checked before each journey to ensure that they are not loose and, at a minimum, each year they should be inspected and replaced if any damage or deterioration is noted. If any of the following issues are identified during an inspection, the mounts should be changed:

- If there are cracks and rust around the metal plate.
- If there is rubber deterioration – rub the rubber and if it leaves a black or sooty residue, the rubber is deteriorating.
- The plate has separated from the rubber – it has 'delaminated'.

Engine bracket showing where a previous failure has been welded. This is due to these parts becoming obsolete on older engines.

To tighten the nuts on the engine mount, always tighten the bottom nuts, never the top one as this adjusts the height and therefore will affect engine alignment.

ENGINE MOUNTING BRACKETS (ENGINE LEGS)

Every engine will have a bracket that allows the engine mounts to be fixed to the engine; this bracket will be specific for your engine model and may be cast aluminium or steel. These brackets are prone to cracking and shearing and should be inspected at least yearly, but if there is a sudden change in engine tone or vibration they should be checked immediately. These components are a regular casualty if you hit an underwater obstacle, or if the engine is out of alignment, or the mounts are failing. As they carry the engine load, this can result in expensive repairs if not picked up and replaced.

RUDDERS AND STEERING

Cruiser: The rudder on a cruiser is usually controlled from the helm using a wheel connected via cables; the operation and maintenance of these are covered in the cable section above. However, it is worth noting that these cables are likely to be fed through the boat and therefore the new cable should always be attached to the old cable and pulled through. Never pull out the cable and then try to feed the new one through the same route, as it can be extremely time-consuming and difficult to undertake. If you have the time, it is worth feeding a second cable along the same route so that in an emergency it can be quickly replaced.

Narrow boat: The rudder is usually held in place with an upper and lower cup containing a bearing. If the rudder is caught on an underwater obstacle, it can come out of the cup and the steering will become loose and 'floppy', making the boat difficult to control. The rudder can be manhandled back into place, usually without getting into the water, although it can be a two-man job. It is simply a case of manoeuvring it into position and you will feel it drop back into the cup. Occasionally, you may need to get into the water to check the bottom cup, as a severe jolt can bend or snap the skeg and this will prevent the rudder placement. The bearing can also

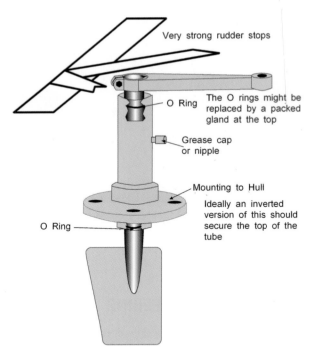

Very strong rudder stops

O Ring

The O rings might be replaced by a packed gland at the top

Grease cap or nipple

Mounting to Hull

Ideally an inverted version of this should secure the top of the tube

O Ring

Typical arrangement of a rudder.

fail and in both cases the vessel may need to be removed from the water in order for repairs to be carried out.

Some rudders pass through the deck at the stern. Occasionally, water may appear bubbling out of the connection due to the pressure from the water. This is not unusual and will not normally cause any

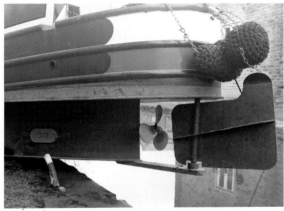

Rudder, showing all components and locations of bearings at the top and bottom of the rudder.

issues. Some rudders have greasing po... greasing will ensure reliability and longevi...

ENGINE SERVICE

In the previous pages, we have covered each ... the systems and how to maintain them. When you undertake a service of the engine you will incorporate the maintenance of all of these systems. A service of the engine involves the following and all additional information needed to complete a service is provided in this section:

- Fuel-filter (including pre-filters) change
- Oil-filter change
- Engine oil change.

Optional:

- Fan-belt change
- Air-filter change
- Gearbox oil change
- Antifreeze top-up
- Stern gland adjustment
- Battery top-up.

AIR FILTER

The air filter is used to filter the air that is needed to allow combustion. Without a clean and adequate supply of air the engine will labour – an analogy would be that when an engine is blocked up, it is similar to trying to breathe with a pillow over your face. Air filters should be regularly checked and replaced or cleaned. There are a number of different types of air filter; most are paper or cloth, but on older engines they may be mesh and simply require cleaning. Changing the air filter is a relatively simple exercise.

When an air filter is blocked, the engine may start to smoke or labour, which may be more evident when the engine is put under load, so that as you increase the revs, power is lost. If your vessel has ever suffered from an exhaust leak, it is always worth checking and changing the air filter, as carbon will block the air filter rapidly.

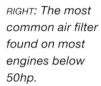

LEFT: Typical air filter from an industrial engine; the air-filter housing will be the same shape.

RIGHT: The most common air filter found on most engines below 50hp.

Air-filter variations; the air-filter housing will be the same shape.

Mesh-type air filter. When unbolted, the mesh can be cleaned by soaking in solvent, paraffin or similar.

Typical air-filter housing found on most engines.

To access the filter, undo the clips around the edge. The filter sits inside and is located on a central raised section inside the housing.

To change an air filter, locate the filter housing, which is usually a large cylindrical metal container with clips around the edge, then simply unclip the lid to gain access to the filter and remove it, replacing it with new. The air filter will normally sit inside the housing without the need to make adjustments; if you experience any difficulty, it may be that the filter is the wrong type.

CHECKING, ADJUSTING OR REPLACING ALTERNATOR/FAN/DRIVE BELT

The alternator belt is also referred to as the fan belt or drive belt and is used to drive the alternator on all engines and in most cases it also drives the engine water pump. The most important element is to ensure that the tension and alignment are correct and these should be regularly checked. If a fan belt is loose, this can cause it to slip. This is usually identifiable by a high-pitched squealing, which will result in early failure of the belt. If the belt is too tight, it can damage the bearings on the alternator, water pump or fan.

A failed belt will have different effects depending upon what it is driving. If it drives the water pump the engine should not be run, as overheating will quickly occur and cause damage. If it is just the alternator, it will stop charging the battery and the ignition light will be illuminated. In these cases, the engine will continue to run and you may only become aware of a problem once the engine refuses to start due to low battery power.

Checking the condition of the belt as part of your maintenance programme is essential to ensuring that breakdowns are kept to a minimum. Fan belts are subject to a harsh environment, as the rubber is stretched and heated (due to friction) each time they are used; they are then cooled and left in a damp environment for long periods. It is important to be able to identify the key signs that indicate a belt is ready for replacement, but in any case the belt should be replaced each year – if it's going to fail, it will fail when you least expect it and in the most inconvenient location, so this should be avoided.

If your fan belt is failing regularly, it is worth getting the alignment and pulley checked. However, if you have a high-output alternator, or have upgraded to one, this can place additional stresses on the belt and swapping to a heavy duty belt, or one rated for high temperatures, may help to resolve this.

Maintenance

To check a V fan belt, first visually inspect it. Look for signs of fraying, cracks and deterioration. The

To check a V-belt tension, push against the belt; there should be no more than 10mm movement.

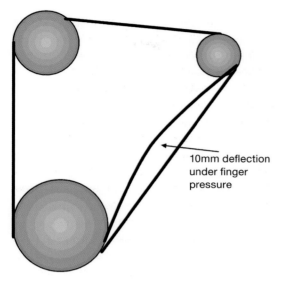

10mm deflection
under finger
pressure

ABOVE: *Attempt to turn the water pump by hand; excessive movement will indicate that the belt needs tightening.*

LEFT: *How a fan belt should be checked and inspected.*

best way to do this is to twist a section of the fan belt between your fingers. Next, check that the tension is correct by pushing against the longest section of the belt. The amount of movement *should* be minor – approximately 10mm. If you are unsure about the tension, locate the water-pump pulley and try to turn it by hand: if there is no movement, the tension is good; if there is movement; the tension needs to be increased.

To check a flat belt, first visually inspect it. Look for signs of fraying, cracks and deterioration. You can also twist the belt; you should feel resistance if it is twisted by more than 90 degrees. Look for worn

edges on the belt, as this indicates that the pulley is wearing the sides. Finally, try to squeeze it between your fingers so that it folds in on itself. There should be resistance and the belt should spring back into position if it is in good condition. If the belt is not tight enough, you will need to adjust the tension, or may need to replace it.

Replacing Alternator/Fan Belts

Replacing the fan belt will depend upon your engine and arrangement. An arrangement with a single alternator is relatively simple and is the same process as for adjusting the belt tension, but a new belt will need to be fed on to the alternator and looped around the pulley before adjusting the tension.

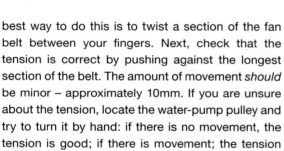

To check a 'flat belt', first visually inspect it. You can also twist the belt and should feel resistance if twisted by more than 90 degrees.

TIP: SPARES

If you have multiple belts on the engine, carrying spares will save you time, but always make sure they are the right size. As an alternative, carrying an adjustable Fenner belt, which is a linked belt, can get you out of trouble. However, it will need tensioning up regularly until it settles in.

The following procedure explains how to adjust and replace a fan belt:

- Slacken off the adjustment arm bolt. This is the one that bolts through the sliding section of the adjustment arm. The alternator should now be able to move, allowing the alternator belt tension to be adjusted (the use of a lever may be required to persuade the alternator to pivot; however, be careful not to damage anything while levering). If the alternator will not move. the 'pivot' mounting bolts may need loosening.
- Slacken off the alternator top 'pivot' mounting bolts, to allow the alternator to be moved. This helps to maintain alignment during tensioning. These bolts allow the alternator to pivot towards or away from the engine when tensioning the belt. They can be located at the top or bottom of the alternator depending on the installation, but will always be on the opposite side to the adjustment arm. Alternators mounted in this way can have one or two mounting points secured by one or two bolts; some with two mounting points use a single bolt (common on Vetus engines).

- Once the fan belt has been fitted (or to adjust) place a levering tool (a hammer shaft or any wooden shaft will do) between the engine and the alternator, so that you can place leverage against the alternator to pull against the belt. Some manufacturers provide a threaded adjuster that moves the alternator along the slot in the adjustment arm, removing the need for the lever.
- Once you have a good tension and whilst keeping the pressure on the belt, tighten the bolts on the adjustment arm and alternator, then recheck the tension.

1. Locate the alternator adjustment arm that is connected to the alternator.

2. Loosen the main tensioning nut.

3. When loose, take hold of the belt and pull upwards. This will move the alternator along the adjustment arm and allow you to remove the belt.

4. Replace the belt and then use a bar or the handle of a hammer or similar and place it behind the alternator to place tension on the alternator.

5. As you apply tension, the alternator will travel along the alternator adjustment arm. When the belt feels at the right tension, use a spanner to tighten the main bolt.

- Check that the top and bottom pulley line up. If they do not, the alternator is misaligned and this will cause early failure of belts and inefficient charging.

On some engines, there may be a number of belts driving starter alternators, domestic alternators and sometimes 240 generators. These can present more of a challenge, as you will need to loosen each alternator in turn and may have to remove belts in order to access others. The process is the same as for a single alternator, but is more time-consuming and complicated. The series of photographs details how to adjust and replace a domestic belt.

1. To adjust the domestic belt, locate the alternator adjustment arm.

2. In this case there are two bolts, one on the slider and one at the top of the arm; loosen the one on the slider first.

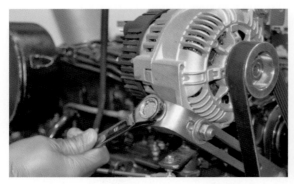

3. Then begin to loosen the top bolt. This releases the tension on the alternator and once finger-loose will allow you to move the alternator. Do not remove the bolt completely.

4. Once loose, grasp the belt and squeeze. This will cause the alternator to move along the slider.

5. Remove the belt by lifting it off the smaller pulley and replace with a new one.

6. Using the top bolt, screw back in until the belt is correctly tensioned, then tighten the bolt on the slider to hold it in position.

Common Problems and Solutions

Misalignment of the pulleys can cause accelerated belt wear. Most misalignment can be improved by using shims of various thickness, or washers between the alternator and the mounting points on the engine.

Squealing is usually a result of a slack belt, but some will squeal on first start regardless of tension. This often occurs when large alternators are recharging batteries first thing in the morning. It

usually stops within a few minutes as alternator load decreases and can be helped by using a proprietary belt dressing. Changing pressed metal pulleys for one machined from solid metal can sometimes help.

Symptoms of belt failure include black dust from the fan belt, in which case check the condition of the pulleys; if they are rusty they will wear the fan-belt edges. If the fan belt fails due to worn edges, check that the belt is the correct width, as if it is too thin it will sit deeper in the groove and the pulley edges will eat away at it.

WINTERIZATION

Winterization is the term that describes the action of preparing a vessel for the winter. It generally involves draining down the raw-water cooling system and, depending upon the arrangement of the cooling system, may require additional steps to protect the engine and vessel adequately.

In the UK, our winters in comparison to many countries are considered mild and therefore as long as there is good antifreeze content in any sealed systems, there is no need to drain down any part of the cooling system unless it is a raw-water cooling system. Raw-water systems on direct raw-water-cooled and heat-exchanger-cooled engines will need to be isolated and drained of all water. It is worth noting that if the system is not winterized and anything happens to the vessel or the engine, you may find that your insurance will be invalidated.

Follow this procedure to drain and refill the raw-water cooling system:

- Close the seacock to isolate the system – as this will be left in this position, it is worth putting a tag on the lever and also on the ignition key to warn 'stopcock isolated'.
- Open the strainer to allow air into the system.
- On a raw-water-cooled system, open all drain valves on the engine.
- On heat-exchanger-cooled engines, remove drain plugs from metal-end caps, or slacken hose clips on rubber-end caps to drain the tube stack in the heat exchanger.

- Remove the raw water from any bends in the pipework – squeeze or blow through to try to remove all water, as any water sat in a bend can potentially freeze.
- Remove the raw-water pump faceplate and let the water drain.
- The majority of hydraulic gearbox oil coolers are plumbed into the sealed system and do not need to be drained, although if yours is in the raw-water circuit it will need to be drained too.

NOTE: If you cruise during winter months this process will need to be carried out each time you leave the boat.

TIP: BEST PRACTICE

Whilst coolant is out of the engine, air is allowed in, which can result in corrosion and rust developing, so the system should be flushed or filled with inhibitor.

Antifreeze

The engine coolant can be pure water, but it is always recommended that either antifreeze or corrosion inhibitor is added to the system. There are many different types of antifreeze and you should always refer to your engine manufacturer's handbook for guidance. Inhibitor can be used instead of antifreeze where the temperature is moderate and freeze protection is not required.

Antifreeze comes in a variety of different colours and can be alcohol or ethylene glycol based. Not all types are equal and will vary in level of protection and temperature ranges. Antifreeze does not degrade significantly over time, but leaks in the cooling system, maintenance and topping up can reduce the antifreeze to water ratio and therefore regular testing and topping up of the antifreeze are advised. Antifreeze corrosion protection, however, does reduce over time and therefore replacement of the antifreeze/coolant mix is required periodically. Normally, the coolant will consist of a fifty/fifty mixture of water and antifreeze. In colder climates,

Thermostat housing showing the corrosion and deterioration that can be caused by not using inhibitor or antifreeze in an engine cooling system.

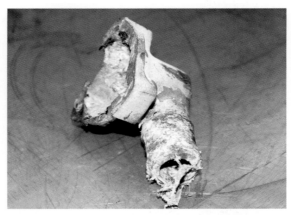

Thermostat showing the pipe entrance blocked due to a build-up of scale.

you may want to increase the antifreeze percentage to give better protection. You can obtain an antifreeze tester from most motor factors.

To drain the system, you will need to identify the drain taps on the engine and the exhaust system. These are usually found at the lowest point on cast components. If your engine is old and this has not been undertaken previously, you will probably find that the drains are seized or blocked. In many engines, there may also be bleed points that will need to be opened to let the circuit empty. If you have a calorifier attached to the engine cooling system, all of the cooling water will need to be drained from this system (and refilled). This can be quite a significant amount, so plan beforehand where or how you are going to store and dispose of the antifreeze mix.

ENGINE DETAILS

Engine Make _____ (enter details)

Engine Model/Serial No. _____ (enter details)

Gearbox Make_____ (enter details)

(Mechanical/Hydraulic) (delete as appropriate)

Gearbox Model/Serial No./Ratio _____ (enter details)

Oil Cooler: Y/N (delete as appropriate)

Cooling System: Raw-Water Cooled/Heat-Exchanger Cooled/Keel Cooled (delete as appropriate)

Electrical System: 12V/24V/other_____(delete as appropriate)

Component Details

Starter Motor Details _____ (enter details)

Alternator (Starter) _____ (enter details)

Alternator (Domestic) _____ (enter details)

Water Pump Type _____ (enter details)

Type of Coupling (Note Shaft Size)_____ (enter details)

Propeller Size _____ (enter details)

Drive Plate No._____ (enter details)

Engine Service Information

Pre-filter: Y/N Details:_____ (enter details)

Fuel Filter No. _____ (enter details)

Oil Filter No. _____ (enter details)

Engine Oil Type_____ (enter details)

Fan Belt Sizes _____ (enter details)

Impeller Details _____ (enter details)

Gearbox Oil Engine Oil/ATF/Other_____ (delete or enter details)

Cable Length & Type _____ (enter details)

SERVICE LOG AND RECORD OF MAINTENANCE UNDERTAKEN

Date	Details of Work	Parts Fitted

GLOSSARY

Agglomerator   Specific form of water trap.

Alternator   Driven by the engine and giving an electrical supply to the boat. Boats use most of the alternator's output for battery charging. If there is just one alternator on the engine it will usually charge both battery banks by a split-charge system.

Aquadrive   *See* Flexible coupling. This joint has two points of flex, so makes engine fitting easy and also is very good at isolating the hull from noise caused by the engine banging the shaft into the bearings. It also contains a thrust block that stops the propeller from pushing and pulling the engine about.

Cable steering   The steering wheel has a drum behind it that moves a long flexible steel cord. The cord runs over a number of pulley wheels to get to the rudder. If the tension on the cord is too great, the system becomes stiff to use. If too slack, the cord falls off the pulleys and jams.

Cardan shaft   Usually found on traditional narrow boats. It is a shaft with two universal joints on it carrying power from the gearbox to propeller shaft. There should be a plummer or thrust block where it joins the propeller shaft.

Centaflex   This maker produces a range of flexible couplings. It is vital to establish the exact model when talking about them.

Chain steering   The steering wheel has a toothed wheel behind it that moves a chain. The chain is used in such a way that it can move the rudder.

Control cable   Transmits movement between the control lever and the gearbox or engine. Most controls use two cables: one for speed; and one for ahead/neutral/astern. It is vital that the clamps holding the ends of the cable in place are kept tight. If they become loose, the cable will not work.

Control or single-lever control   A mechanism that allows the boater to use a single lever to control both the boat's gearbox and engine speed. There are many makes and types.

Cutless bearing   *See* Shaft bearing.

Domestic alternator   Alternator on a twin- or two-alternator engine that charges the domestic or house battery bank.

Drive plate   Disc that is bolted to the engine flywheel and which then drives the gearbox. When this starts to fail, there is often more noise and when it totally fails the boat is likely to lose all drive. There are many different designs and makes.

ECM   *See* ECU.

ECU   Computer on petrol-injection systems. Large diesels by the likes of Volvo may control the diesel system with an ECU. Also controls the ignition system on similar modern petrol engines.

Engine alternator Alternator on a twin- or two-alternator engine that charges the engine battery.

Engine foot/leg/bracket Bracket that is the part of the engine and/or gearbox that fixes the engine into the boat.

Engine mount or flexible mount Component that fixes the engine foot to the boat containing a block of rubber that supports the engine.

Flexible coupling Engines that are fitted on flexible mounts tend to jump about a bit, so to allow for that a flexible coupling is usually fitted between the gearbox and the shaft. There are lots of different makes and designs. Some common makers are Centaflex, R&D, Aquadrive and Python-Drive. Some boats even use a shaft with joints, like old cars used to use. These are known as universal joints, or UJs.

Flexibly mounted engine Engine mounted on flexible mounts.

Flywheel Large, heavy wheel at the back of the engine that spins around all the time the engine is running. You should never see this item, because it is inside a housing at the back of the engine known as the flywheel housing.

Fuel filter Normally only found on diesel engines, but some petrol engines use them as water traps. Cleans the fuel, so that no dirt can damage the injector pump and injectors.

Fuel line Metal or rubber tube that carries fuel to where it is needed.

Fuel pump Pump that is either driven by the engine or electricity. It pumps fuel from the fuel tank to the engine.

Fuel tank Big 'tin' that holds the fuel.

Gearbox Gives neutral, ahead and astern gears. On inboard boats, it is bolted to the back of the engine. On outboards, it is in the part of the engine that is below the water. On Z drives/outdrives, it is in the part that is hung off the back of the boat. Outdrives and outboards have gearboxes internally. Some common manufacturers are: PRM, Hurth, TMP, Technodrive, Lister (on older engines).

Hydraulic drive System that transmits power from the engine to the shaft by means of an oil pump, pipes and a hydraulic motor. Uses similar parts to JCBs and mobile cranes. If the system runs out of oil or a pipe bursts all drive is lost.

Hydraulic gearbox Also *see* Gearbox. Gearboxes where the boater only moves a valve and oil pressure actually changes the gear. The valve is moved via a control cable. Common makes: larger PRM, Velvet Drive, TMP, Lister LH150. These boxes lose drive when they run short of oil, but the LH150 locks into ahead.

Hydraulic ram Sealed tube with a rod and piston in it and two pipe connections. The rod passes through one end cap of the tube. The piston and rod are forced in one direction or the other, depending upon which pipe is forcing oil into the ram.

Hydraulic steering The steering wheel has an oil pump and reservoir behind it. Oil is pumped through a pair of pipes that link the pump to a hydraulic ram, which in turn moves the rudder. The direction the oil flows in depends upon which way the wheel is turned, which in turn either forces the ram open or closed.

Inboard engine Engine that is fitted inside the boat.

Injector Part on a diesel engine that squirts fuel into the engine.

Injector pipe   Pipe that carries fuel from the injector pump to the injector on a diesel engine.

Injector pump   Another pump that is only found on diesel engines. It accurately measures the fuel and pressurizes it to more than 2,000psi and delivers it to the right injector.

Jack shaft   *See* Cardan shaft.

Lift pump   *See* Fuel pump. Usually refers to the fuel pump used on a diesel engine.

Mechanical gearbox   Also *see* Gearbox. Gearboxes where the moving parts are moved by the boater, usually via a control cable, but on older boats by a big lever. These will still drive when short of oil, but may become noisy and the gears may be hard to select. Common makes are Hurth, the smaller PRMs and Lister 100.

Mechanical steering   Includes the chain, cable and push-pull cable steering, but also refers to a number of systems that use small gearboxes, sliding rods or rotating shafts to pass the movement from the wheel to the rudder.

Morse cable   *See* Control cable.

Morse control   *See* Control or single-lever control.

Outboard motor   Engine that is hung on the outside of the boat. Usually a petrol engine, but there are a few diesel ones about.

Outdrive   Another name for a Z drive. The engine is inside the boat, but the propeller and gearbox are part of a mechanism outside the boat. This mechanism is a bit like the bottom half of an outboard.

Plummer block   See Thrust block or bearing.

Prop-shaft or propeller shaft   Long metal rod that runs from the back of the gearbox, through the bottom of the boat to the propeller. This is what drives the propeller.

Push-pull cable   The ordinary control cable is a push-pull cable, but on steering systems this cable is far larger and stronger. It is vital that the clamps holding the ends of the cable in place are kept tight. If they become loose, the boat will lose steering.

Python-Drive   *See* Flexible couplings and Aquadrive. Very similar to an Aquadrive, but a different make.

Rigidly or solidly mounted engine   Engine bolted straight into the boat with no flexible mounts. Usually on lengths of solid wood.

Sail drive   Inboard engine with what is in effect an outdrive fitted through the bottom of the boat. This is mainly found on yachts.

Sedimentor   Specific form of water trap.

Shaft bearing   The things that support the shaft. On most narrow boats they are a plain brass/bronze tube, but on cruisers and yachts they are likely to be a fluted rubber sleeve in a brass tube. These are known as Cutless bearings.

Skeg   An extension of the keel to allow the rudder to be mounted.

Split-charge diode   Method for charge-splitting using diodes.

Split-charge relay   Method for charge-splitting using an electrically operated switch known as a relay.

Split-charge switch   Large, multi-position switch that gives manual control of which battery bank is being charged or used.

Split-charge system Allows one alternator to charge two battery banks, but prevents one bank discharging itself into the other bank.

Starter motor or starter Small but powerful electric motor used to start an engine turning.

Stern gland Type of seal to prevent water leaking into the boat where the shaft goes through the hull. Most need daily attention to grease them and adjustment, especially if they start to leak. Increasingly, glands that need little attention are being fitted. Some common names of this type are Vetus, Volvo, Deep Sea, PSS. All of these look as if they are made of nearly all rubber.

Teleflex control *See* Control or single-lever control.

Thrust block or bearing Supports the propeller shaft and usually transmits thrust to the boat's hull.

Tiller steering Simple method of directing the boat using a long arm fixed to the rudder mechanism. Usually found on narrow boats, yachts and a few workboats.

V drive Inboard engine and gearbox, except the engine is fitted to the boat backwards and the gearbox or another box turns the propeller shaft back, so that it runs under the engine and out of the boat. This allows the engine to be placed very close to the back of the boat.

Water trap Filter-like object located between the fuel tank and the engine, whose job is to remove water and some dirt from the fuel. May be found on both inboard petrol and diesel boats.

Wheel steering The boater uses a wheel to direct the boat. This is the most common type of steering on motor boats and larger yachts. The boater's movement is transmitted by a variety of mechanisms, such as a push-pull cable, wire rope/chain, or a hydraulic system.

Z drive Another name for an outdrive.

INDEX

RELATED TITLES FROM CROWOOD

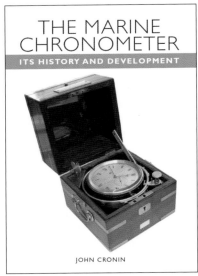

The Marine Chronometer
JOHN CRONIN
ISBN 978 1 84797 185 2
150 illustrations, 112pp

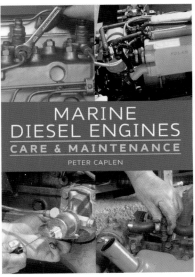

Marine Diesel Engines
PETER CAPLEN
ISBN 978 1 84797 175 3
370 illustrations, 192pp

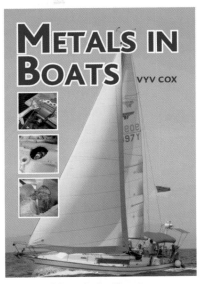

Metals in Boats
VYV COX
ISBN 978 1 78500 262 5
110 illustrations, 128pp

Narrow Boats
NICK BILLINGHAM
ISBN 978 1 85223 861 2
110 illustrations, 144pp

In case of difficulty ordering, please contact the Sales Office:

The Crowood Press, Ramsbury, Wiltshire SN8 2HR UK

Tel: 44 (0) 1672 520320   enquiries@crowood.com   www.crowood.com